Nobel Laureates
in
Physics

1901–2011

Nobel Laureates in Physics

1901–2011

Sathish LA MSc PhD

Assistant Professor and coordinator
Postgraduate Department of Physics
Government Science College
(NAAC Re Accredited A Grade)
Nrupatunga Road, Bangalore
Karnataka, India

CBS

CBS Publishers & Distributors Pvt Ltd

New Delhi • Bengaluru • Pune • Kochi • Chennai

Nobel Laureates in
Physics
1901–2011

ISBN: 978-81-239-2211-9

First Edition: 2012

Published by Satish Kumar Jain and produced by Vinod K. Jain for
CBS Publishers & Distributors Pvt Ltd
4819/XI Prahlad Street, 24 Ansari Road, Daryaganj, New Delhi 110 002, India.

Website: www.cbspd.com
Ph: 23289259, 23266861, 23266867   e-mail: delhi@cbspd.com
Fax: 011-23243014   publishing@cbspd.com
    cbspubs@airtelmail.in

Corporate Office: 204 FIE, Industrial Area, Patparganj, Delhi 110 092
Ph: 4934 4934   Fax: 4934 4935

Branches

- **Bengaluru:** Seema House 2975, 17th Cross, K.R. Road, Banasankari 2nd Stage, Bengaluru 560 070, Karnataka
  Ph: +91-80-26771678/79   Fax: +91-80-26771680   e-mail: bangalore@cbspd.com
- **Pune:** Bhuruk Prestige, Sr. No. 52/12/2+1+3/2 Narhe, Haveli (Near Katraj-Dehu Road Bypass), Pune 411 041, Maharashtra
  Ph: +91-20-64704058, 64704059, 32342277   Fax: +91-020-24300160   e-mail: pune@cbspd.com
- **Kochi:** 36/14 Kalluvilakam, Lissie Hospital Road, Kochi 682 018, Kerala
  Ph: +91-484-4059061-65   Fax: +91-484-4059065   e-mail: cochin@cbspd.com
- **Chennai:** 20, West Park Road, Shenoy Nagar, Chennai 600 030, Tamil Nadu
  Ph: +91-44-26260666, 26208620   Fax: +91-44-45530020   email: chennai@cbspd.com

Printed at Magic International Pvt Ltd., Greater Noida

*One and all who
have contributed
to the growth
of physics*

Preface

Writing a book on Nobel Prizes in physics, the Nobel laureates, and their works is not an easy task. My intention is to sketch a broad outline on the scientists who have won the Nobel Prizes in physics. This book gives a concise information on their works (however, the given information should not be judged by the number of pages devoted to it).

The purpose of this book is to inspire the young minds and also all those who read physics. This is an attempt to enlighten the heritage of physics. The students, teachers and all the readers of scientific community will appreciate the convenience of a single volume.

The present volume covers the early philosophers of physics and their works in brief till the inception of Nobel Prize in physics that highlights the foundation work in physics, and brief information about Alfred Nobel and the initiation of Nobel Foundation.

The book too includes the portraits of scientists who have crowned the Nobel Prizes in physics since 1901 and their works in brief with 210 photographs.

In spite of my best efforts, there could still be some errors left uncorrected. I shall be grateful to the readers who find and report mistakes and deficiencies in this volume.

Sathish LA

email: lasgayit@yahoo.com

Acknowledgements

At the outset I would like to bow my head to the Almighty for bestowing me the intellectual power while writing this volume.

I wish to thank all the Principals of Government Science College of Bangalore for their incessant support at all stages. Special thanks are due to Mr SK Jain, Managing Director, Mr YN Arjuna, Senior Publishing Director, New Delhi, and Mr Deepak Rao, Bangalore, for having confidence in me to compile this unique volume.

I would like to place on record my deep sense of gratitude to Mr YN Arjuna for a candid review of the book which considerably helped me in improving the materials presented.

I have drawn heavily from the available literature on the subject to compile this book which is gratefully acknowledged; I owe my debt to all those pieces of literature which I have read on the legacy of physics.

The photographs are the courtesy of http://en.wikipedia.org/wiki/Nobel Prize in Physics, which is highly acknowledged.

I also wish to express my sincere gratitude to all those who have worked on this book, especially the MAP System, Bangalore, for the good quality DTP work.

Words are not enough to express the support given by my wife and daughter at all times.

Sathish LA
email: lasgayit@yahoo.com

Contents

NOBEL PRIZE IN PHYSICS

Thales of Miletus

 Thales was born in the city of Miletus. Miletus was an ancient Greek Ionian city on the western coast of Asia Minor (Aydin Province of Turkey), near the Maeander River.

Born	ca. 624–625 BC
Died	ca. 547–546 BC
School	Ionian philosophy, Milesian school, naturalism
Main interests	Ethics, metaphysics, mathematics, astronomy
Notable ideas	Water is the physis, Thales' theorem
Influenced by	Babylonian astronomy and ancient Egyptian mathematics and religion
Influenced	Pythagoras, Anaximander, Anaximenes

The dates of Thales' life are not known precisely. The time of his life is roughly established by a few dateable events mentioned in the sources and an estimate of his length of life. According to Herodotus, Thales once predicted a solar eclipse which has been determined by modern methods to have been on 28 May, 585 BC. Diogenes Laërtius quotes the chronicle of Apollodorus as saying that Thales died at 78 in the 58th Olympiad (548–545), and Sosicrates as reporting that he was 90 at his death.

Diogenes Laërtius states that (according to Herodotus and Douris and Democritus) his parents were Examyes and Cleobuline, Phoenician nobles. Giving another opinion, he ultimately connects Thales' family line back to Phoenician prince Cadmus. Diogenes also reports two other stories, one that he married and had a son, Cybisthus or Cybisthon, or adopted his nephew of the same name. The second is that he never married, telling his mother as a young man that it was too early to marry, and as an older man that it was too late. A much earlier source, Plutarch, tells the following story: Solon who visited Thales asked him the reason which kept him single. Thales answered that he did not like the idea of having to worry about children. Nevertheless, several years later Thales, anxious for family, adopted his nephew Cybisthus.

Thales involved himself in many activities, taking the role of an innovator. Some say that he left no writings, others that he wrote 'On the Solstice' and 'On the Equinox'. Neither has survived. Diogenes Laërtius quotes letters of Thales to Pherecydes and Solon, offering to review the book of the former on religion, and offering to keep company with the latter on his sojourn from Athens. Thales identifies the Milesians as Athenians.

Thales of Miletus was a pre-Socratic Greek philosopher from Miletus in Asia Minor, and one of the Seven Sages of Greece. Many, most notably Aristotle, regard him as the first philosopher in the Greek tradition. According to Bertrand Russell, Western philosophy begins with Thales. Thales attempted to explain natural phenomena without reference to mythology and was tremendously influential in this respect. Almost all the other pre-Socratic philosophers follow him in attempting to provide an explanation of ultimate substance, change, and the existence of the world without reference to mythology. Those philosophers were also influential, and eventually Thales' rejection of mythological explanations became an essential idea for the scientific revolution. He was also the first to define general principles and set forth hypotheses, and as a result has been dubbed the Father of Science.

Pythagoras

Pythagoras of Samos was an Ionian Greek philosopher and founder of the religious movement called Pythagoreanism. Most of our information about Pythagoras was written down centuries after he lived, thus very little reliable information is known about him.

Born	c. 570 BC, Samos Island
Died	c. 495 BC (age around 75), Metapontum
Era	Ancient philosophy
Region	Western philosophy
School	Pythagoreanism
Main interests	Metaphysics, music, mathematics, ethics, politics
Notable ideas	Musica universalis, Golden ratio, Pythagorean tuning, Pythagorean theorem
Influenced by	Thales, Anaximander, Pherecydes
Influenced	Philolaus, Alcmaeon, Parmenides, Plato, Euclid, Empedocles, Hippasus, Kepler

He was born on the island of Samos, and may have travelled widely in his youth, visiting Egypt and other places seeking knowledge. Around 530 BC, he moved to Croton, a Greek colony in southern Italy, and there set up a religious sect. His followers pursued the religious rites and practices developed by Pythagoras, and studied his philosophical theories. The society took an active role in the politics of Croton, but this eventually led to their downfall. The Pythagorean meeting-places were burned, and Pythagoras was forced to flee the city. He is said to have ended his days in Metapontum.

Pythagoras made influential contributions to philosophy and religious teaching in the late 6th century BC. He is often revered as a great mathematician, mystic and scientist, and is best known for the Pythagorean Theorem. However, because legend and obfuscation cloud his work even more than with the other pre-Socratic philosophers, one can say little with confidence about his teachings, and some have questioned whether he contributed much to mathematics and natural philosophy. Many of the accomplishments credited to Pythagoras may actually have been accomplishments of his colleagues and successors. Whether or not his disciples believed that everything was related to mathematics and that numbers were the ultimate reality which is unknown. It was said that he was the first man to call himself a philosopher, or lover of wisdom, and Pythagorean ideas exercised a marked influence on Plato.

Accurate facts about the life of Pythagoras are very few, and most information concerning him is of late date, and so untrustworthy, that it is impossible to provide more than a vague outline of his life. The lack of information by contemporary writers, together with the secrecy which surrounded the Pythagorean brotherhood, meant that invention took the place of facts. The stories which were created, were eagerly sought by the Neoplatonist writers who provide most of the details about Pythagoras, but who were uncritical concerning anything which were related to the Gods or which was considered divine. Thus many myths were created – such as that Apollo was his father; that Pythagoras gleamed with a supernatural brightness; that he had a golden thigh; that Abaris came flying to him on a golden arrow; that he was seen in different places at one and the same time.

With the exception of a few remarks by Xenophanes, Heraclitus, Herodotus, Plato, Aristotle, and Isocrates, we are mainly dependent on Diogenes Laërtius, Porphyry, and Iamblichus for the biographical details.

Hence historians are often reduced in considering the statements based on their inherent probability, but even then, if all the credible stories concerning Pythagoras were supposed to be true, his range of activity would be impossibly vast.

Anaxagoras

Anaxagoras (lord of the assembly) was a Pre-Socratic Greek philosopher. Born in Clazomenae in Asia Minor, Anaxagoras was the first philosopher to bring philosophy from Ionia to Athens. He attempted to give a scientific account of eclipses, meteors, rainbows, and the sun, which he described as a fiery mass larger than the Peloponnese. He was accused of contravening the established religion and was forced to flee to Lampsacus.

Born	c. 500 BC, Clazomenae
Died	c. 428 BC (age around 72), Lampsacus
Era	Ancient philosophy
Region	Western philosophy
School	Pluralist school
Main interests	Natural philosophy
Notable ideas	Cosmic mind (Nous) ordering all things
Influenced by	Milesian school
Influenced	Archelaus

Anaxagoras is famous for introducing the cosmological concept of Nous (mind), as an ordering force. He regarded material, substance as an infinite multitude of imperishable primary elements, referring all generation and disappearance to mixture and separation respectively.

Anaxagoras appears to have had some amount of property and prospects of political influence in his native town of Clazomenae in Asia Minor. However, he supposedly surrendered both of these out of a fear that they would hinder his search for knowledge. Valerius Maximus preserves a different tradition: Anaxagoras, coming home from a long voyage, found his property in ruin, and said "If this had not perished, I would have".This is a sentence that says the Roman denoting the most perfect wisdom. Although as a Greek, he may have been a soldier of the Persian army when Clazomenae was suppressed during the Ionian Revolt.

In early manhood (c. 464–461 BC) he went to Athens, which was rapidly becoming the centre of Greek culture. There he is said to have remained for thirty years. Pericles learned to love and admire him, and the poet Euripides derived from him an enthusiasm for science and humanity.

Anaxagoras brought philosophy and the spirit of scientific inquiry from Ionia to Athens. His observations of the celestial bodies and the fall of meteorites led him to form new theories of the universal order. He was the first to explain that the Moon shines due to reflected light from the Sun. He also said that the Moon had mountains and he believed that it was inhabited. The heavenly bodies, he asserted, were masses of stone torn from the earth and ignited by rapid rotation. He explained that though both Sun and the stars were fiery stones, we do not feel the heat of the stars because of their enormous distance from Earth. He thought that the Earth is flat and floats supported by strong air under it and disturbances in this air sometimes cause earthquakes. However, these theories brought him into collision with the popular faith; Anaxagoras views on such things as heavenly bodies were considered dangerous.

At about 450 BC, Anaxagoras was arrested by Pericles political opponents on a charge of contravening the established religion. It took Pericles power of persuasion to secure his release. Even so he was forced to retire from Athens to Lampsacus in Troad (c. 434–433 BC). He died there in around the year 428 BC. Citizens of Lampsacus erected an altar to Mind and Truth in his memory, and observed the anniversary of his death for many years. Anaxagoras wrote a book of philosophy, but only fragments of the first part of this have survived, through preservation in work of Simplicius of Cilicia in the sixth century AD.

Empedocles

 Empedocles was a Greek pre-Socratic philosopher and a citizen of Agrigentum, a Greek city in Sicily. Empedocles philosophy is best known for being the origin of the cosmogenic theory of the four Classical elements.

Born	490 BC, Agrigentum, Sicily
Died	430 BC (age around 60), Mount Etna, Sicily
Era	Pre-Socratic philosophy
Region	Western philosophy
School	Pluralist school
Main interests	Cosmogenesis and Ontology
Notable ideas	All matter is made up of four elements: water, earth, air and fire
Influenced by	Parmenides, Pythagoreanism
Influenced	Gorgias of Leontini, Aristotle, Lucretius, Friedrich Nietzsche

He also proposed powers called Love and Strife which would act as forces to bring about the mixture and separation of the elements. These physical speculations were part of a history of the universe, which also dealt with the origin and development of life. Influenced by the Pythagoreans, he supported the doctrine of reincarnation. Empedocles is generally considered the last Greek philosopher to record his ideas in verse. Some of his work still survives today, more so than in the case of any other Pre-Socratic philosopher. Empedocles death was mythologized by ancient writers, and has been the subject of number of literary treatments.

Empedocles was born in c. 490 BC, at Agrigentum in Sicily to a distinguished family. Very little is known about his life. His father Meto seems to have been instrumental in overthrowing the tyrant of Agrigentum, presumably Thrasydaeus in 470 BC. Empedocles continued the democratic tradition of his house by helping to overthrow the succeeding oligarchic government. He is said to have been magnanimous in his support of the poor; severe in persecuting the overbearing conduct of the Aristocrats; and he even declined the sovereignty of the city when it was offered to him.

His brilliant oratory, his penetrating knowledge of nature, and the reputation of his marvellous powers, including the curing of diseases, and averting epidemics, produced many myths and stories surrounding his name. He was said to have been a magician and controller of storms, and he himself, in his famous poem Purifications seems to have promised miraculous powers, including the destruction of evil, the curing of old age, and the controlling of wind and rain.

Empedocles was acquainted or connected by friendship with the physicians Acron and Pausanias, who was his eromenos; with various Pythagoreans; and even, it is said, with Parmenides and Anaxagoras. The only pupil of Empedocles who is mentioned is the sophist and rhetorician Gorgias.

Timaeus and Dicaearchus spoke of the journey of Empedocles to the Peloponnese, and of the admiration which was paid to him there; others mentioned his stay at Athens, and in the newly-founded colony of Thurii, 446 BC. there are also fanciful reports of him travelling far to the east to the lands of the Magi.

According to Aristotle, he died at the age of sixty (c. 430 BC), even though other writers have him living up to the age of one hundred and nine. Likewise, there are myths concerning his death: a tradition, which is traced to Heraclides Ponticus, represented him as having been removed from the earth; whereas others had him perishing in the flames of Mount Etna.

Democritus

Democritus (chosen of the people) was an Ancient Greek philosopher born in Abdera, Thrace, Greece. He was an influential pre-Socratic philosopher and pupil of Leucippus, who formulated an atomic theory for the cosmos.

Born	ca. 470–460 BC, Abdera, Thrace
Died	ca. 360–370 BC (age 90–109)
Era	Pre-Socratic philosophy
Region	Western philosophy
School	Pre-Socratic philosophy
Main interests	Metaphysics, mathematics, astronomy
Notable ideas	Atomism, Distant Star Theory
Influenced by	Leucippus, Melissus of Samos
Influenced	Epicurus, Pyrrho, Lucretius, Santayana, Aristotle

His exact contributions are difficult to disentangle from his mentor Leucippus, as they are often mentioned together in texts. Their speculation on atoms, taken from Leucippus, bears a passing and partial resemblance to the 19th century, understanding of atomic structure that has led some to regard Democritus as more of a scientist than other Greek philosophers; nevertheless their ideas rested on very different bases. Largely ignored in Athens, Democritus was nevertheless well-known to his fellow northern-born philosopher Aristotle. Plato is said to dislike him so much that he wished all his books burnt. Many consider Democritus to be the father of modern science.

Although some called him a Milesian. His year of birth was 460 BC according to Apollodorus, who is probably more reliable than Thrasyllus who placed it ten years earlier. John Burnet has argued that the date of 460 is too early, since according to Diogenes Laërtius ix.41, Democritus said that he was a young man (neos) during Anaxagoras old age (circa 440–428). It was said that Democritus father was so wealthy that he received Xerxes on his march through Abdera. Democritus spent the inheritance which his father left him on travels into distant countries, to satisfy his thirst for knowledge. He travelled to Asia, and was even said to have reached India and Ethiopia. We know that he wrote on Babylon and Meroe; he must also have visited Egypt, and Diodorus Siculus states that he lived there for five years. He himself declared that among his contemporaries none had made greater journeys, seen more countries, and met more scholars than himself. He particularly mentions the Egyptian mathematicians, whose knowledge he praises. Theophrastus, too, spoke of him as a man who had seen many countries. During his travels, according to Diogenes Laërtius, he became acquainted with the Chaldean magi. A certain Ostanes, one of the magi accompanying Xerxes was also said to have taught him. After returning to his native land he occupied himself with natural philosophy. He traveled throughout Greece to acquire a knowledge of its culture. He mentions many Greek philosophers in his writings, and his wealth enabled him to purchase their writings. Leucippus, the founder of the atomism, was the greatest influence upon him. He also praises Anaxagoras. Diogenes Laërtius says that he was friends with Hippocrates. He may have been acquainted with Socrates, but Plato does not mention him and Democritus himself is quoted as saying, "I came to Athens and no one knew me". Aristotle placed him among the pre-Socratic natural philosophers.

Aristotle

Aristotle was a Greek philosopher, a student of Plato and teacher of Alexander the Great. His writings cover many subjects, including physics, metaphysics, poetry, theater, music, logic, rhetoric, politics, government, ethics, biology, and zoology. Together with Plato and Socrates (Plato's teacher), Aristotle is one of the most important founding figures in Western philosophy.

Born	384 BC, Stageira, Chalcidice
Died	322 BC (age 61 or 62), Euboea
Era	Ancient philosophy
Region	Western philosophy
School	Peripatetic school, Aristotelianism
Main interests	Physics, metaphysics, poetry, theatre, music, rhetoric, politics, government, ethics, biology, zoology
Notable ideas	Golden mean, reason, logic, passion
Influenced by	Parmenides, Socrates, Plato, Heraclitus, Democritus
Influenced	Virtually all Western philosophy after his works, Alexander the Great, Avicenna, Averroes, Maimonides, Albertus Magnus, Thomas Aquinas, Duns Scotus, Ptolemy, Copernicus, Galileo, and most of Islamic philosophy, Jewish philosophy, Christian philosophy, science and more

Aristotle's writings constitute first at creating a comprehensive system of Western philosophy, encompassing morality and aesthetics, logic and science, politics and metaphysics. Aristotle's views on the physical sciences profoundly shaped medieval scholarship, and their influence extended well into the Renaissance, although they were ultimately replaced by

Newtonian physics. In the biological sciences, some of his observations were confirmed to be accurate only in the 19th century. His works contain the earliest known formal study of logic, which was incorporated in the late 19th century into modern formal logic. In metaphysics, Aristotelianism had a profound influence on philosophical and theological thinking in the Islamic and Jewish traditions in the Middle Ages, and it continues to influence Christian theology, especially Eastern Orthodox theology, and the scholastic tradition of the Catholic Church. His ethics, though always influential, gained renewed interest with the modern advent of virtue ethics. All aspects of Aristotle's philosophy continue to be the object of active academic study today. Though Aristotle wrote many elegant treatises and dialogues (Cicero described his literary style as a river of gold), it is thought that the majority of his writings are now lost and only about one-third of the original works have survived. Despite the far-reaching appeal that Aristotle's works have traditionally enjoyed, today modern scholarship questions a substantial portion of the Aristotelian corpus as authentically Aristotle's own.

Aristotle was born in Stageira, Chalcidice, in 384 BC, about 55 km east of modern-day Thessaloniki. His father Nicomachus was the personal physician to King Amyntas of Macedon. Aristotle was trained and educated as a member of the aristocracy. At about the age of eighteen, he went to Athens to continue his education at Plato's Academy. Aristotle remained at the academy for nearly twenty years, not leaving until after Plato's death in 347 BC. He then traveled with Xenocrates to the court of his friend Hermias of Atarneus in Asia Minor. While in Asia, Aristotle traveled with Theophrastus to the island of Lesbos, where together they researched the botany and zoology of the island. Aristotle married Hermias's adoptive daughter (or niece) Pythias. She bore him a daughter, whom they named Pythias. Soon after Hermias' death, Aristotle was invited by Philip II of Macedon to become the tutor of his son Alexander the Great in 343 BC.

Aristotle was appointed as the head of the royal academy of Macedon. During that time he gave lessons not only to

Alexander, but also to two other future kings: Ptolemy and Cassander. In his Politics, Aristotle states that only one thing could justify monarchy, and that was the virtue of the king and his family were greater than the virtue of the rest of the citizens put together. Tactfully, he included the young prince and his father in that category. Aristotle encouraged Alexander toward Eastern conquest, and his attitude towards Persia was unabashedly ethnocentric. In one famous example, he counsels Alexander to be a leader to the Greeks and a despot to the barbarians, to look after the former as after friends and relatives, and to deal with the latter as with beasts or plants.

By 335 BC he had returned to Athens, establishing his own school there known as the Lyceum. Aristotle conducted courses at the school for the next twelve years. While in Athens, his wife Pythias died and Aristotle became involved with Herpyllis of Stageira, who bore him a son whom he named after his father, Nicomachus. According to the Suda, he also had an eromenos, Palaephatus of Abydus.

It is during this period in Athens from 335 to 323 BC when Aristotle is believed to have composed many of his works. Aristotle wrote many dialogues, only fragments of which survived. The works that have survived are in treatise form and were not, for the most part, intended for widespread publication, as they are generally thought to be lecture aids for his students. His most important treatises include physics, metaphysics, Nicomachean ethics, politics, De Anima (on the soul) and poetics.

Aristotle not only studied almost every subject possible at the time, but made significant contributions to most of them. In physical science, Aristotle studied anatomy, astronomy, embryology, geography, geology, meteorology, physics and zoology. In philosophy, he wrote on aesthetics, ethics, government, metaphysics, politics, economics, psychology, rhetoric and theology. He also studied education, foreign customs, literature and poetry. His combined works constitute a virtual encyclopaedia of Greek knowledge. It has been suggested that Aristotle was probably the last person to know everything that was to be known in his own time.

Near the end of Alexander's life, Alexander began to suspect plots against himself, and threatened Aristotle in letters. Aristotle had made no secret of his contempt for Alexander's pretense of divinity, and the king had executed Aristotle's grandnephew Callisthenes as a traitor. A widespread tradition in antiquity suspected Aristotle of playing a role in Alexander's death, but there is little evidence for this.

Upon Alexander's death, anti-Macedonian sentiment in Athens once again flared. Eurymedon the hierophant denounced Aristotle for not holding the Gods in honor. Aristotle fled the city to his mother's family estate in Chalcis, explaining, I will not allow the Athenians to sin twice against philosophy, a reference to Athens's prior trial and execution of Socrates. However, he died in Euboea of natural causes within the year (in 322 BC). Aristotle named chief executor his student Antipater and left a will in which he asked to be buried next to his wife.

Aristotle says that on the subject of reasoning he had nothing else on an earlier date to speak of. However, Plato reports that syntax was devised before him, by Prodicus of Ceos, who was concerned by the correct use of words. Logic seems to have emerged from dialectics; the earlier philosophers made frequent use of concepts like reductio ad absurdum in their discussions, but never truly understood the logical implications. Even Plato had difficulties with logic; although he had a reasonable conception of a deducting system, he could never actually construct one and relied instead on his dialectic. Plato believed that deduction would simply follow from premises, hence he focused on maintaining solid premises so that the conclusion would logically follow. Consequently, Plato realized that a method for obtaining conclusions would be most beneficial. He never succeeded in devising such a method, but his best attempt was published in his book Sophist, where he introduced his division method.

Aristarchus of Samos

 Aristarchus was a Greek astronomer and mathematician, born on the island of Samos, in Greece. He is the first known person to have presented a heliocentric model of the solar system, placing the Sun, not the Earth, at the centre of the known universe.

Born	310 BC in Greece
Died	230 BC in Greece
Birth place	Samos Greece
Race	White
Occupation	Astronomer
Nationality	Ancient Greece

He was influenced by the Pythagorean Philolaus of Croton, but, in contrast to Philolaus, he had both identified the central fire with the Sun, as well as put other planets in correct order from the Sun. His astronomical ideas were often rejected in favor of the geocentric theories of Aristotle and Ptolemy until they were successfully revived nearly 1800 years later by Copernicus and extensively developed and built upon by Johannes Kepler and Isaac Newton. The crater Aristarchus on the Moon is named in his honor.

The only work usually attributed to Aristarchus that has survived to the present time, On the Sizes and Distances of the Sun and Moon, is based on a geocentric world view. It is peculiar and possibly informative that this work reckons the Sun's diameter as 2°, rather than the correct value, 1/2°. The latter diameter is known from Archimedes to have been Aristarchus's actual value.

Though the original text has been lost, a reference in Archimedes book The Sand Reckoner (Archimedis Syracusani

Arenarius & Dimensio Circuli) describes another work by Aristarchus in which he advanced an alternative hypothesis of the heliocentric model.

Archimedes wrote: You (King Gelon) are aware, the universe is the name given by most astronomers to the sphere the center of which is the center of the Earth, while its radius is equal to the straight line between the center of the Sun and the center of the Earth. This is the common account as you have heard from astronomers. But Aristarchus has brought out a book consisting of certain hypotheses, wherein it appears, as a consequence of the assumptions made, that the universe is many times greater than the universe just mentioned. His hypotheses are that the stars are fixed and the Sun remain unmoved, that the Earth revolves around the Sun on the circumference of a circle, the Sun lying in the middle of the Floor, and that the sphere of the fixed stars, situated about the same center as the Sun, is so great that the circle in which he supposes the Earth to revolve bears such a proportion to the distance of the fixed stars as the center of the sphere bears to its surface.

Aristarchus thus believed the stars to be very far away, and saw this as the reason why there was no visible parallax, that is an observed movement of the stars relative to each other as the Earth moved around the Sun. The stars are in fact much farther away than the distance that was generally assumed in ancient times, which is why stellar parallax is only detectable with telescopes. The geocentric model, consistent with planetary parallax, was assumed to be an explanation for the unobservability of the parallel phenomenon, stellar parallax. The rejection of the heliocentric view was common, as the following passage from Plutarch suggests (on the apparent face in the orb of the Moon):

Cleanthes (a contemporary of Aristarchus and head of the Stoics) thought it was the duty of the Greeks to indict Aristarchus on the charge of impiety for putting in motion the earth of the universe supposing the heaven to remain at rest and the earth to revolve in an oblique circle, while it rotates, at the same time, about its own axis.

Archimedes

Archimedes of Syracuse was a Greek mathematician, physicist, engineer, inventor and astronomer. Although few details of his life are known, he is regarded as one of the leading scientists in classical antiquity. Among his advances in physics are the foundations of hydrostatics, statics and an explanation of the principle of the lever.

Born	c. 287 BC, Syracuse, Sicily, Magna Graecia
Died	c. 212 BC (age around 75), Syracuse
Residence	Syracuse, Sicily
Fields	Mathematics, Physics, Engineering, Astronomy, Invention
Known for	Archimedes Principle, Archimedes screw, Hydrostatics, Levers, Infinitesimals
Archimedes is said to have remarked about the lever: Give me a place to stand on and I will move the Earth.	

He is credited with designing innovative machines, including siege engines and the screw pump that bears his name. Modern experiments have tested claims that Archimedes designed machines capable of lifting attacking ships out of the water and setting ships on fire using an array of mirrors.

Archimedes is generally considered to be the greatest mathematician of antiquity and one of the greatest of all time. He used the method of exhaustion to calculate the area under the arc of a parabola with the summation of an infinite series, and gave a remarkably accurate approximation of pi. He also defined the spiral bearing his name, formulae for the volumes of surfaces of revolution and an ingenious system for expressing very large numbers.

Archimedes died during the Siege of Syracuse when he was killed by a Roman soldier despite the orders that he should not be harmed. Cicero describes visiting the tomb of Archimedes, which was surmounted by a sphere inscribed within a cylinder. Archimedes had proven that the sphere has two-thirds of the volume and surface area of the cylinder (including the bases of the latter), and regarded this as the greatest of his mathematical achievements.

Unlike his inventions, the mathematical writings of Archimedes were little known in antiquity. Mathematicians from Alexandria read and quoted him, but the first comprehensive compilation was not made until *c.* 530 AD by Isidore of Miletus, while commentaries on the works of Archimedes written by Eutocius in the 6th century AD opened them to wider readership for the first time. The relatively few copies of Archimedes written work that survived through the Middle Ages were an influential source of ideas for scientists during the Renaissance, while the discovery in 1906 of previously unknown works by Archimedes in the Archimedes Palimpsest has provided new insights into how he obtained mathematical results.

Archimedes died in c. 212 BC during the Second Punic War, when Roman forces under General Marcus Claudius Marcellus captured the city of Syracuse after a two year long siege. According to the popular account given by Plutarch, Archimedes was contemplating a mathematical diagram when the city was captured. A Roman soldier commanded him to come and meet General Marcellus but he declined, saying that he had to finish working on the problem. The soldier was enraged by this, and killed Archimedes with his sword. Plutarch also gives a lesser-known account of the death of Archimedes which suggests that he may have been killed while attempting to surrender to a Roman soldier. According to this story, Archimedes was carrying mathematical instruments, and was killed because the soldier thought that they were valuable items. General Marcellus was reportedly angered by the death of Archimedes, as he considered him a valuable scientific asset and had ordered that he was not be harmed.

Nicolaus Copernicus

 Nicolaus Copernicus was a Renaissance astronomer and the first person to formulate a comprehensive heliocentric cosmology, which displaced the Earth from the center of the universe.

Born	19 February 1473, Toru (Thorn), Royal Prussia, Kingdom of Poland
Died	24 May 1543 (age 70), Frombork (Frauenburg), Prince-Bishopric of Warmia, Kingdom of Poland
Fields	Mathematics, astronomy, Canon law, medicine, economics
Alma mater	Kraków University, Bologna University, University of Padua, University of Ferrara
Known for	Heliocentrism

Copernicus epochal book, De Revolutionibusbium coelestium (On the Revolutions of the Celestial Spheres), was published just before his death in 1543, is often regarded as the starting point of modern astronomy and the defining epiphany that began the scientific revolution. His heliocentric model, with the Sun at the center of the universe, demonstrated that the observed motions of celestial objects can be explained without putting Earth at rest in the center of the universe. His work stimulated further scientific investigations, becoming a landmark in the history of science that is often referred to as the Copernican Revolution.

Among the great polymaths of the Renaissance, Copernicus was a mathematician, astronomer, physician, quadrilingual polyglot, classical scholar, translator, artist, Catholic cleric, jurist, governor, military leader, diplomat and economist. Among his many responsibilities, astronomy figured as little more than an avocation, yet it was in that field that he made his mark upon the world.

Copernicus is said to be the founder of modern astronomy. He was born in Poland and eventually was sent off to Cracow University, to study mathematics and optics; at Bologna, Canon law. Returning from his studies in Italy, Copernicus, through the influence of his uncle, was appointed as a canon in the cathedral of Frauenburg where he spent a sheltered and academic life for the rest of his days. Because of his clerical position, Copernicus moved in the highest circles of power; but he remained a student. For relaxation Copernicus painted and translated Greek poetry into Latin. His interest in astronomy gradually grew to be one in which he had a primary interest. His investigations were carried on quietly and alone, without help or consultation. He made his celestial observations from a turret situated on the protective wall around the cathedral, observations were made bare eyeball, so to speak, as a hundred more years were to pass before the invention of the telescope. In 1530, Copernicus completed and gave the world his great work De Revolutionibus, which asserted that the earth rotated on its axis once daily and traveled around the sun once yearly, a fantastic concept for the times. Up to the time of Copernicus the thinkers of the western world believed in the Ptolemiac theory that the universe was a closed space bounded by a spherical envelope beyond which there was nothing. Claudius Ptolemy, an Egyptian living in Alexandria, at about 150 AD gathered and organized the thoughts of the earlier thinkers. Ptolemy's findings were that the earth was fixed, inert, immovable mass, located at the center of the universe, and all celestial bodies, including the sun and the fixed stars, revolved around it. It was a theory that appealed to human nature. It fit with the casual observations that a person might want to make in the field; and second, it fed man's ego.

Having completed all his studies in Italy, 30-year-old Copernicus returned to Warmia, where apart from brief journeys to Kraków and to nearby Prussian cities (Toruń, Gdańsk, Elbląg, Grudziądz, Malbork, Königsberg) — he would live out the remaining 40 years of his life.

Galileo Galilei

Galileo Galilei was an Italian physicist, mathe-matician, astronomer and philosopher who played a major role in the Scientific Revolution. His achievements include improvements to the telescope and consequent astronomical obser-vations, and support for Copernicanism.

Born	15 February 1564, Pisa, Duchy of Florence, Italy
Died	8 January 1642 (age 77), Arcetri, Grand Duchy of Tuscany, Italy
Residence	Grand Duchy of Tuscany, Italy
Nationality	Italian
Fields	Astronomy, physics and mathematics
Institutions	University of Pisa, University of Padua
Alma mater	University of Pisa
Academic advisors	Ostilio Ricci
Notable students	Benedetto Castelli, Mario Guiducci, Vincenzio Viviani
Known for	Kinematics, dynamics, telescopic observational astronomy heliocentrism

Galileo has been called the father of modern observational astronomy, the father of modern physics, the father of science, and the Father of Modern Science. Stephen Hawking says, Galileo, perhaps more than any other single person, was responsible for the birth of modern science.

The motion of uniformly accelerated objects, taught in nearly all high school and introductory college physics courses, was studied by Galileo as the subject of kinematics. His contributions to observational astronomy include the telescopic confirmation of the phases of Venus, the discovery of the four largest satellites of Jupiter (named the Galilean Moons in his honour), and the observation and analysis of Sunspots. Galileo also worked in applied science and technology, inventing an improved military compass and other instruments.

Galileo's championing of Copernicanism was controversial within his lifetime, when a large majority of philosophers and astronomers still subscribed to the geocentric view that the Earth is at the centre of the universe. After 1610, when he began publicly supporting the heliocentric view, which placed the Sun at the centre of the universe, he met with bitter opposition from some philosophers and clerics, and two of the latter eventually denounced him to the Roman Inquisition early in 1615. In February 1616, although he had been cleared of any offence, the Catholic Church nevertheless condemned heliocentrism as false and contrary to Scripture, and Galileo was warned to abandon his support for it—which he promised to do. When he later defended his views in his most famous work, Dialogue Concerning the Two Chief World Systems, published in 1632, he was tried by the Inquisition, found vehemently suspect of heresy, forced to recant, and spent the rest of his life under house arrest.

In 1619, Galileo became embroiled in a controversy with Father Orazio Grassi, professor of mathematics at the Jesuit Collegio Romano. It began as a dispute over the nature of comets, but by the time Galileo had published The Assayer (Il Saggiatore) in 1623, his last salvo in the dispute, it had become a much wider argument over the very nature of science itself. Because The Assayer contains such a wealth of Galileo's ideas on how science should be practiced, it has been referred to as his scientific manifesto.

Tycho Brahe

Tycho Brahe was a Danish nobleman known for his accurate and comprehensive astronomical and planetary observations. Coming from Scania, then part of Denmark, now part of modern-day Sweden, Tycho was well known in his lifetime as an astronomer and alchemist.

Born	14 December 1546, Knutstorp Castle, Scania
Died	24 October 1601 (age 54), Prague
Nationality	Danish
Education	Private
Occupation	Nobleman, astronomer
Religion	Lutheran

He adopted the Latinized name Tycho Brahe from *Tycho* at around age fifteen, and he is now generally referred to as Tycho as was common in Scandinavia in his time, rather than by his surname Brahe (The incorrect form of his name, *Tycho de Brahe*, appeared only much later.)

Tycho Brahe was granted an estate on the island of Hven and the funding to build the Uraniborg, an early research institute, where he built large astronomical instruments and took many careful measurements. After disagreements with the new king in 1597, he was invited by the Bohemian king and Holy Roman emperor Rudolph II to Prague, where he became the official imperial astronomer. He built the new observatory at Benátky nad Jizerou. Here, from 1600 until his death in 1601, he was assisted by Johannes Kepler. Kepler later used Tycho's astronomical information to develop his own theories of astronomy.

As an astronomer, Tycho worked to combine what he saw as the geometrical benefits of the Copernican system with the philosophical benefits of the Ptolemaic system into his own model of the universe, the Tychonic system.

Tycho is credited with the most accurate astronomical observations of his time and the data was used by his assistant Kepler to derive the laws of planetary motion. No one before Tycho had attempted to make so many planetary observations.

On 11 November 1572, Tycho observed a very bright star, now named SN 1572, which had unexpectedly appeared in the constellation Cassiopeia. Because it had been maintained since antiquity that the world beyond the Moon's orbit was eternally unchangeable (celestial immutability was a fundamental axiom of the Aristotelian world-view), other observers held that the phenomenon was something in the terrestrial sphere below the Moon. However, in the first instance Tycho observed that the object showed no daily parallax against the background of the fixed stars. This implied it was at least farther away than the Moon and those planets that do show such parallax. If the object was closer to the Earth than the Moon, it would be observed to move against a fixed background of stars, as the Moon is observed to do. However, the object showed no such motion and appeared to be at least further than the Moon. Moreover he also found the object did not even change its position relative to the fixed stars over several months as all planets did in their periodic orbital motions, even the outer planets for which no daily parallax was detectable. This suggested it was not even a planet, but a fixed star in the stellar sphere beyond all the planets. In 1573 he published a small book, De nova stella thereby coining the term nova for a new star. This discovery was decisive for his choice of astronomy as a profession. Tycho was strongly critical of those who dismissed the implications of the astronomical appearance, writing in the preface to De nova stella: "O crassa ingenia. O caecos coeli spectatores" (Oh thick wits. Oh blind watchers of the sky).

Johannes Kepler

Johannes Kepler was a German mathematician, astronomer and astrologer, and key figure in the 17th century scientific revolution. He is best known for his eponymous laws of planetary motion, codified by later astronomers based on his works Astronomia nova, Harmonices Mundi, and Epitome of Copernican Astronomy. They also provided one of the foundations for Isaac Newton's theory of universal gravitation.

Born	27 December 1571, Weil der Stadt near Stuttgart, Germany
Died	15 November 1630 (age 58), Regensburg, Bavaria, Germany
Residence	Württemberg, Styria, Bohemia, Upper Austria
Fields	Astronomy, astrology, mathematics and natural philosophy
Institutions	University of Linz
Alma mater	University of Tübingen
Known for	Kepler's laws of planetary motion, Kepler conjecture

During his career, Kepler was a mathematics teacher at a seminary school in Graz, Austria, an assistant to astronomer Tycho Brahe, the imperial mathematician to Emperor Rudolf II and his two successors Matthias and Ferdinand II, a mathematics teacher in Linz, Austria, and an adviser to General Wallenstein. He also did fundamental work in the field of optics, invented an improved version of the refracting telescope and helped to legitimize the telescopic discoveries of his contemporary Galileo Galilei.

Kepler lived in an era when there was no clear distinction between astronomy and astrology, but there was a strong division between astronomy (a branch of mathematics within the liberal arts) and physics (a branch of natural philosophy). Kepler also incorporated religious arguments and reasoning into his work, motivated by the religious conviction that God had created the world according to an intelligible plan that is accessible through the natural light of reason. Kepler described his new astronomy as celestial physics, as an excursion into Aristotle's Metaphysics, and as a supplement to Aristotle's On the Heavens, transforming the ancient tradition of physical cosmology by treating astronomy as part of a universal mathematical physics.

Johannes Kepler's first major astronomical work, Mysterium Cosmographicum (The Cosmographic Mystery), was the first published defense of the Copernican system. Kepler claimed to have had an epiphany on 19 July, 1595, while teaching in Graz, demonstrating the periodic conjunction of Saturn and Jupiter in the zodiac; he realized that regular polygons bound one inscribed and one circumscribed circle at definite ratios, which, he reasoned, might be the geometrical basis of the universe. After failing to find a unique arrangement of polygons that fit known astronomical observations, Kepler began experimenting with 3-dimensional polyhedra. He found that each of the five Platonic solids could be uniquely inscribed and circumscribed by spherical orbs; nesting these solids, each encased in a sphere, within one another would produce six layers, corresponding to the six known planets Mercury, Venus, Earth, Mars, Jupiter, and Saturn. By ordering the solids correctly octahedron, icosahedron, dodecahedron, tetrahedron, cube Kepler found that the spheres could be placed at intervals corresponding to the relative sizes of each planet's path, assuming the planets circle the Sun. Kepler also found a formula relating the size of each planet's orb to the length of its orbital period: from inner to outer planets, the ratio of increase in orbital period is twice the difference in orb radius. However, Kepler later rejected this formula, because it was not precise enough.

Sir Isaac Newton

Sir Isaac Newton was an English physicist, mathematician, astronomer, natural philosopher, alchemist, and theologian who is considered by many scholars and members of the general public to be one of the most influential people in human history.

Born	4 January 1643, [OS: 25 December 1642] Woolsthorpe-by-Colsterworth Lincolnshire, England
Died	31 March 1727 (age 84), [OS: 20 March 1726], Kensington, Middlesex, England
Residence	England
Fields	Physics, mathematics, astronomy, natural philosophy, alchemy, Christiaan theology
Institutions	University of Cambridge, Royal Society, Royal Mint
Alma mater	Trinity College, Cambridge
Academic advisors	Isaac Barrow, Benjamin Pulleyn
Notable students	Roger Cotes, William Whiston
Known for	Newtonian mechanics, universal gravitation, infinitesimal calculus optics, binomial series, Newton's method, Philosophiae Naturalis Principia Mathematica
Influences	Henry More, Polish Brethren
Influenced	Nicolas Fatio de Duillier, John Keill

His 1687 publication of the Philosophiae Naturalis Principia Mathematica (usually called the Principia) is considered to be among the most influential books in the history of science, laying the ground work for most of classical mechanics. In this work, Newton described universal gravitation and the three laws of motion which dominated the scientific view of the physical universe for the next three centuries. Newton showed that the motions of objects on Earth and of celestial bodies are governed by the same set of natural laws by demonstrating the consistency between Kepler's laws of planetary motion and his theory of gravitation, thus removing the last doubts about heliocentrism and advancing the scientific revolution.

Newton built the first practical of reflecting telescope and developed a theory of colour based on the observation, that a prism decomposes white light into many colours that form the visible spectrum. He also formulated an empirical law of cooling and studied the speed of sound.

In mathematics, Newton shares the credit with Gottfried Leibniz for the development of the differential and integral calculus. He also demonstrated the generalised binomial theorem, developed Newton's method for approximating the roots of a function, and contributed to the study of power series.

Newton remains uniquely influential to scientists, as demonstrated by a 2005 survey of members of Britain's Royal Society asking who had the greater effect on the history of science and had the greater contribution to humankind, Newton or Albert Einstein. Royal Society scientists deemed Newton to have made the greater overall contribution on both.

Newton's work has been said to distinctly advance every branch of mathematics then studied. His work on the subject usually referred to as fluxions or calculus is seen, for example, in a manuscript of October 1666, now published among Newton's mathematical papers. A related subject was infinite series. Newton's manuscript De analysi per aequationes numero terminorum infinitas (on analysis by equations infinite in number of terms) was sent by Isaac Barrow to John Collins in June 1669: in August 1669 Barrow identified its author to Collins as Mr Newton, a fellow of our College, and very young ... but of an extraordinary genius and proficiency in these things.

Christiaan Huygens

Christiaan Huygens, was a prominent Dutch mathematician, astronomer, physicist, horologist, and writer of early science fiction.

Born	14 April 1629, The Hague, Netherlands
Died	8 July 1695 (age 66), Netherlands
Residence	Netherlands, France
Nationality	Dutch
Fields	Physics, mathematics, astronomy, horology, science fiction
Institutions	Royal Society of London, French Academy of Sciences
Alma mater	University of Leiden, College of Orange
Academic advisors	Frans van Schooten, John Pell
Known for	Titan explanation Saturn's rings, centrifugal force, collision formulae pendulum clock, Huygens–Fresnel principle, wave theory, Birefringence first theoretical physicist
Influences	René Descartes, Frans van Schooten, Blaise Pascal, Marin Mersenne
Influenced	Gottfried Wilhelm Leibniz, Isaac Newton

His work included early telescopic studies elucidating the nature of the rings of Saturn and the discovery of its Moon Titan, the invention of the pendulum clock and other investigations in timekeeping, and studies of both optics and the centrifugal force.

He was born in April 1629 at The Hague, the second son of Constantijn Huygens, (1596–1687), a friend of mathematician and philosopher René Descartes, and of Suzanna van Baerle (deceased 1637), whom Constantijn had married on 6 April 1627. Christiaan studied law and mathematics at the University of Leiden and the College of Orange in Breda. After a stint as a diplomat, Huygens turned to science.

Huygens achieved note for his argument that light consists of waves, now known as the Huygens-Fresnel principle, which became instrumental in the understanding of wave-particle duality. He generally receives credit for his discovery of the centrifugal force, the laws for collision of bodies, for his role in the development of modern calculus and his original observations on sound perception. Huygens is seen as the first theoretical physicist as he was the first to use formulae in physics.

Huygens is remembered especially for his wave theory of light, expounded in his Treatise on light, 1678. The later theory of light by Isaac Newton in his Opticks proposed a different explanation for reflection, refraction and interference of light assuming the existence of light particles. The interference experiments of Thomas Young vindicated Huygens' wave theory in 1801, as the results could no longer be explained with light particles.

Huygens formulated what is now known as the second law of motion of Isaac Newton in a quadratic form. Newton reformulated and generalized that law. In 1659 Huygens derived the now well-known formula for the centripetal force, exerted by an object describing a circular motion, for instance on the string to which it is attached, in modern notation

$Fc = \dfrac{mv^2}{r}$, with m the mass of the object, v the velocity and r the radius.

Furthermore, Huygens concluded that Descartes' laws for the elastic collision of two bodies must be wrong and formulated the correct laws.

Thomas Young

 Thomas Young was an English genius and polymath, admired by, among others, Herschel and Einstein. He is famous for having partly deciphered Egyptian hieroglyphs (specifically the Rosetta Stone) before Jean-Francois Champollion eventually expanded on his work.

Born	13 June 1773, Milverton, Somerset, England
Died	10 May 1829 (age 55), London, England
Fields	Physics, physiology, egyptology

Young made notable scientific contributions to the fields of vision, light, solid mechanics, energy, physiology, language, musical harmony and egyptology. Young belonged to a Quaker family of Milverton, Somerset, he was born in 1773, the eldest of ten children. At the age of fourteen Young had learned Greek and Latin and was acquainted with French, Italian, Hebrew, German, Chaldean, Syriac, Samaritan, Arabic, Persian, Turkish and Amharic.

Young began to study medicine in London in 1792, moved to Edinburgh in 1794, and a year later went to Göttingen, Lower Saxony, Germany where he obtained the degree of doctor of physics in 1796. In 1797 he entered Emmanuel College, Cambridge. In the same year he inherited the estate of his granduncle, Richard Brocklesby, which made him financially independent, and in 1799 he established himself as a physician at 48 Welbeck Street, London (now recorded with a blue plaque). Young published many of his first academic articles anonymously to protect his reputation as a physician.

In 1801 Young was appointed professor of natural philosophy (mainly physics) at the Royal Institution. In two years he delivered 91 lectures. In 1802, he was appointed foreign secretary of the Royal Society, of which he had been elected a fellow in 1794. He resigned his professorship in 1803, fearing that its duties would interfere with his medical practice. His lectures were published in 1807 in the Course of Lectures on Natural Philosophy and contain a number of anticipations of later theories.

In 1811 Young became physician at St. George's Hospital, and in 1814 he served on a committee appointed to consider the dangers involved in the general introduction of gas into London. In 1816 he was secretary of a commission charged with ascertaining the precise length of the second's or seconds pendulum (the length of a pendulum whose period is exactly 2 seconds), and in 1818 he became secretary to the Board of Longitude and superintendent of the HM Nautical Almanac Office.

A few years before his death he became interested in life assurance, and in 1827 he was chosen one of the eight foreign associates of the French Academy of Sciences. In 1828, he was elected as a foreign member of the Royal Swedish Academy of Sciences.

Thomas Young died in London on 10 May 1829, and was buried in the cemetery of St. Giles Church in Farnborough, Kent, England. Later scholars and scientists have praised Young's work, although they may know him only through achievements he made in their fields. His contemporary Sir John Herschel called him a truly original genius. Albert Einstein praised him in the 1931 foreword to an edition of Newton's Opticks. Other admirers include physicist Lord Rayleigh and Nobel laureate Philip Anderson.

Thomas Young's name has been adopted as the name of the London-based Thomas Young Centre, an alliance of academic research groups engaged in the theory and simulation of materials.

Ludwig Boltzmann

 Ludwig Eduard Boltzmann was an Austrian physicist famous for his founding, contributions in the fields of statistical mechanics and statistical thermodynamics. He was one of the most important advocates for the atomic theory when that scientific model was still highly controversial.

Born	20 February 1844, Vienna, Austrian Empire
Died	5 September 1906 (age 62), Duino near Trieste, Italy (at that time Austria-Hungary)
Residence	Austria, Germany
Nationality	Austrian
Fields	Physicist
Institutions	University of Graz, University of Vienna, University of Munich, University of Leipzig
Alma mater	University of Vienna
Doctoral advisor	Josef Stefan
Doctoral students	Paul Ehrenfest, Philipp Frank, Gustav Herglotz, Franc Hocevar Ignacij Klemencic, Lise Meitner
Known for	Boltzmann's constant, Boltzmann equation, H-theorem, Boltzmann distribution, Stefan-Boltzmann law

Boltzmann was born in Vienna, the capital of the Austrian Empire. His father, Ludwig Georg Boltzmann, was a tax official. His grandfather, who had moved to Vienna from Berlin, was a clock manufacturer, and Boltzmann's mother, Katharina Pauernfeind, was originally from Salzburg. He received his primary education from a private tutor at the home of his parents. Boltzmann attended high school in Linz, Upper Austria.

Boltzmann studied physics at the University of Vienna, starting in 1863. Among his teachers were Josef Loschmidt, Joseph Stefan, Andreas von Ettingshausen and Jozef Petzval. Boltzmann received his PhD degree in 1866 working under the supervision of Stefan; his dissertation was on kinetic theory of gases. In 1867 he became a Privatdozent (lecturer). After obtaining his doctorate degree, Boltzmann worked two more years as Stefan's assistant. It was Stefan who introduced Boltzmann to Maxwell's work.

In 1869, at age 25, he was appointed full Professor of Mathematical Physics at the University of Graz in the province of Styria. In 1869 he spent several months in Heidelberg working with Robert Bunsen and Leo Königsberger and then in 1871 he was with Gustav Kirchhoff and Hermann von Helmholtz in Berlin. In 1873 Boltzmann joined the University of Vienna as Professor of Mathematics and there he stayed until 1876.

In 1872, long before women were admitted to Austrian universities, he met Henriette von Aigentler, an aspiring teacher of mathematics and physics in Graz. She was refused permission to unofficially audit lectures, and Boltzmann advised her to appeal; she did, successfully. On 17 July, 1876 Ludwig Boltzmann married Henriette; they had three daughters and two sons. Boltzmann went back to Graz to take up the chair of Experimental Physics. Among his students in Graz were Svante Arrhenius and Walther Nernst. He spent 14 happy years in Graz and it was there that he developed his statistical concept of nature. In 1885 he became a member of the Imperial Austrian Academy of Sciences and in 1887 he became the President of the University of Graz. He was elected a member of the Royal Swedish Academy of Sciences in 1888.

Boltzmann was appointed to the Chair of Theoretical Physics at the University of Munich in Bavaria, Germany in 1890. In 1893, Boltzmann succeeded his teacher Joseph Stefan as Professor of Theoretical Physics at the University of Vienna.

Thomas Edison

 Thomas Alva Edison was an American inventor, scientist, and businessman who developed many devices that greatly influenced life around the world, including the phonograph, the motion picture camera, and a long-lasting, practical electric light bulb.

Born	11 February 1847, Milan, Ohio
Died	18 October 1931 (age 84), West Orange, New Jersey
Occupation	Inventor, Scientist, Businessman
Religion	Deist
Spouse(s)	Mary Stilwell (m. 1871–1884) , Mina Miller (m. 1886–1931)
Children	Marion Estelle Edison (1873–1965), Thomas Alva Edison Jr. (1876–1935), William Leslie Edison (1878–1937), Madeleine Edison (1888–1979), Charles Edison (1890–1969), Theodore Miller Edison (1898–1992)
Parents	Samuel Ogden Edison, Jr. (1804–1896), Nancy Matthews Elliott (1810–1871)

Genius is one percent inspiration, ninety-nine percent perspiration – Thomas Alva Edison, Harper's Monthly (September 1932)

Dubbed The Wizard of Menlo Park (now Edison, New Jersey) by a newspaper reporter, he was one of the first inventors to apply the principles of mass production and large teamwork to the process of invention, and therefore, is often credited with the creation of the first industrial research laboratory.

Edison is considered one of the most prolific inventors in history, holding 1,093 U.S. patents in his name, as well as many patents in the United Kingdom, France, and Germany. He is credited with numerous inventions that contributed to mass communication and, in particular, telecommunications. His advanced work in these fields was an outgrowth of his early career as a telegraph operator. Edison originated the concept and implementation of electric-power generation and distribution to homes, businesses, and factories–a crucial development in the modern industrialized world. His first power station was at Manhattan Island, New York.

Building on the contributions of other developers over the previous three quarters of a century, Edison made significant improvements to the idea of incandescent light, and wound up in the public consciousness as the inventor of the light bulb.

After many experiments with platinum and other metal filaments, Edison returned to a carbon filament. The first successful test was on 22 October, 1879; it lasted 40 hours. Edison continued to improve this design and by 4 November, 1879, filed for U.S. patent 223,898 (granted on 27 January, 1880) for an electric lamp using a carbon filament or strip coiled and connected to platina contact wires. Although the patent described several ways of creating the carbon filament including cotton and linen thread, wood splints, papers coiled in various ways it was not until several months after the patent was granted that Edison and his team discovered a carbonized bamboo filament that could last over 1,200 hours. The idea of using this particular raw material originated from Edison's recalling his examination of a few threads from a bamboo fishing pole while relaxing on the shore of Battle Lake in the present-day state of Wyoming, where he and other members of a scientific team had traveled so that they could clearly observe a total eclipse of the sun on 29 July, 1878, from the Continental Divide.

In 1878, Edison formed the Edison Electric Light Company in New York City with several financiers, including J. P. Morgan and the members of the Vanderbilt family. Edison made the first public demonstration of his incandescent light bulb on 31 December, 1879, in Menlo Park. It was during this time that he said: We will make electricity so cheap that only the rich will burn candles.

Mahen Theatre in Brno in what is now the Czech Republic, was the first public building in the world to use Edison's electric lamps, with the installation supervised by Edison's assistant in the invention of the lamp, Francis Jehl.

Alfred Nobel

Alfred Bernhard Nobel (Stockholm, Sweden, 21 October 1833 – Sanremo, Italy, 10 December 1896) was a Swedish chemist, engineer, innovator, armaments manufacturer and the inventor of dynamite. He owned Bofors, a major armaments manufacturer, which he had redirected from its previous role as an iron and steel mill. Nobel held 355 different patents, dynamite being the most famous. In his last will, he used his enormous fortune to institute the Nobel Prizes. The synthetic element nobelium was named after him.

Born	21 October 1833, Stockholm, Sweden
Died	10 December 1896 (age 63), Sanremo, Italy
Residence	Norra begravningsplatsen, Stockholm
Occupation	Chemist, engineer, innovator, armaments manufacturer and inventor
Known for	Invention of dynamite, Nobel Prize

He was the third son of Immanuel Nobel and Andriette Ahlsell Nobel. Born in Stockholm on 21 October 1833, he went with his family to Saint Petersburg in 1842, where his father (who had invented modern plywood) started a torpedo works. Alfred studied chemistry with Professor Nikolay Nikolaevich Zinin. When Alfred was 18, he went to the United States to study chemistry for four years and worked for a short period under John Ericsson. In 1859, the factory was left to the care of the second son, Ludvig Nobel, who greatly improved the business. Alfred, returning to Sweden with his father after the bankruptcy of their family business, devoted himself to the study of explosives, and especially to the safe manufacture and use of nitroglycerine (discovered in 1847 by Ascanio Sobrero, one of his fellow students under Théophile-Jules Pelouze at the University of Torino). A big explosion occurred on the 3 September 1864 at their factory in Heleneborg in Stockholm, Killing five people, among them was Alfred's younger brother Emil.

The foundations of the Nobel Prize were laid in 1895 when Alfred Nobel wrote his last will, leaving much of his wealth for its establishment. Since 1901, the prize has honored men and women for outstanding achievements in physics, chemistry, medicine, literature, for work in peace and now economics.

Sri Kantha has suggested that the one personal trait of Nobel that helped him to sharpen his creativity include his talent for information access, via his multi-lingual skills. Despite the lack of formal secondary and tertiary level education, Nobel gained proficiency in six languages: Swedish, French, Russian, English, German and Italian. He also developed literary skills to write poetry in English. His Nemesis, a prose tragedy in four acts about Beatrice Cenci, partly inspired by Percy Bysshe Shelley's The Cenci, was printed while he was dying. The entire stock except for three copies was destroyed immediately after his death, being regarded as scandalous and blasphemous. The first surviving edition (bilingual Swedish–Esperanto) was published in Sweden in 2003. The play has been translated to Slovenian via the Esperanto version. In 2010 it was published in Russia as another bilingual (Russian–Esperanto) edition.

Nobel was elected a member of the Royal Swedish Academy of Sciences in 1884, the same institution that would later select laureates for two of the Nobel Prizes, and he received an honorary doctorate from Uppsala University in 1893. Alfred Nobel is buried in Norra begravningsplatsen in Stockholm.

Nobel found that when nitroglycerin was incorporated in an absorbent inert substance like *kieselguhr* (diatomaceous earth) it became safer and more convenient to handle, and this mixture he patented in 1867 as dynamite. Nobel demonstrated his explosive for the first time that year, at a quarry in Redhill, Surrey, England. In order to help reestablish his name and improve the image of his business from the earlier controversies associated with the dangerous explosives, Nobel had also considered naming the highly powerful substance Nobels

Safety Powder, but settled with dynamite instead, referring to the Greek word for power.

Nobel later on combined nitroglycerin with various nitro-cellulose compounds, similar to collodion, but settled on a more efficient recipe combining another nitrate explosive, and obtained a transparent, jelly-like substance, which was a more powerful explosive than dynamite. Gelignite, or blasting gelatin, as it was named, was patented in 1876; and was followed by a host of similar combinations, modified by the addition of potassium nitrate and various other substances. Gelignite was more stable, transportable and conveniently formed to fit into bored holes, like those used in drilling and mining, than the previously used compounds and was adopted as the standard technology for mining in the Age of Engineering bringing Nobel a great amount of financial success, though at a significant cost to his health.

In 1888 Alfred's brother Ludvig died while visiting Cannes and a French newspaper erroneously published Alfred's obituary. It condemned him for his invention of dynamite and is said to have brought about his decision to leave a better legacy after his death. The obituary stated Le marchand de la mort est mort (The merchant of death is dead) and went on to say, Dr. Alfred Nobel, who became rich by finding ways to kill more people faster than ever before, died yesterday. On 27 November 1895, at the Swedish-Norwegian Club in Paris, Nobel signed his last will and testament and set aside the bulk of his estate to establish the Nobel Prizes, to be awarded annually without distinction of nationality. He died of a stroke on 10 December 1896 at Sanremo, Italy. After taxes and bequests to individuals, Nobel's will gave 31,225,000 Swedish kronor (equivalent to about 1.8 billion kronor or 250 million US dollars in 2008) to fund the prizes.

The first three of these prizes are awarded for eminence in physical science, chemistry and in medical science or physiology; the fourth is for literary work in an ideal direction and the fifth prize is to be given to the person or society that renders the greatest service to the cause of international fraternity, in the suppression or reduction of standing armies, or in the establishment or furtherance of peace congresses.

The Formulation about the literary prize, in an ideal direction (*idealisk riktning* in Swedish), is cryptic and has caused much confusion. For many years, the Swedish Academy interpreted ideal as idealistic (*idealistisk*) and used it as a reason not to give the prize to important but less Romantic authors, such as Henrik Ibsen and Leo Tolstoy. This interpretation has since been revised, and the prize has been awarded to, for example, Dario Fo and José Saramago, who definitely do not belong to the camp of literary idealism.

There was also quite a lot of room for interpretation by the bodies he had named for deciding on the physical sciences and chemistry prizes, given that he had not consulted them before making the will. In his one-page testament, he stipulated that the money go to discoveries or inventions in the physical sciences and to discoveries or improvements in chemistry. He had opened the door to technological awards, but had not left instructions on how to deal with the distinction between science and technology. Since the deciding bodies he had chosen were more concerned with the former, it is not surprising that the prizes went to scientists and not to engineers, technicians or other inventors.

NOBEL PRIZE

The Nobel Prizes are annual international awards bestowed by Scandinavian Committees in recognition of cultural and scientific advances. They were established in 1895 by the Swedish chemist Alfred Nobel, the inventor of dynamite. The prizes in Physics, Chemistry, Physiology or Medicine, Literature, and Peace were first awarded in 1901. The Sveriges Riksbank Prize in Economic Sciences in Memory of Alfred Nobel was instituted by Sveriges Riksbank in 1968 and first awarded in 1969. Although this is not technically a Nobel Prize, its announcements and presentations are made along with the other prizes. Each Nobel Prize is widely regarded as the most prestigious award in its field.

The Royal Swedish Academy of Sciences awards the Nobel Prize in Physics, the Nobel Prize in Chemistry, and the Nobel

Memorial Prize in Economic Sciences. The Nobel Assembly at Karolinska Institute awards the Nobel Prize in Physiology or Medicine and the Swedish Academy grants the Nobel Prize in Literature. The Nobel Peace Prize is not awarded by a Swedish organisation but by the Norwegian Nobel Committee.

Each recipient, or laureate, is presented with a gold medal, a diploma, and a sum of money which depends on the Nobel Foundation's income that year. In 2009, each prize was worth 10 million SEK (c. US $1.4 million). The prize can not be awarded posthumously, nor may a prize be shared among more than three people. These strict rules have deprived worthy nominees of an award.

The members of the Norwegian Nobel Committee that were to award the Peace Prize were appointed shortly after the will was approved. The other prize-awarding organisations followed: the Karolinska Institute on 7 June, the Swedish Academy on 9 June, and the Royal Swedish Academy of Sciences on 11 June. The Nobel Foundation reached an agreement on guidelines for how the prizes should be awarded. In 1900, the Nobel Foundation's newly created statutes were promulgated by King Oscar II. In 1905, the Union between Sweden and Norway was dissolved, which meant the responsibility for awarding Nobel Prizes was split between the two countries. Norway's Nobel Committee became responsible for awarding the Nobel Peace Prize and Swedish institutions remained responsible for the other prizes.

NOBEL FOUNDATION

The Nobel Foundation was founded as a private organization on 29 June 1900, to manage the finances and administration of the Nobel Prizes. In accordance with Nobel's will, the primary task of the Foundation is to manage the fortune Nobel left. Another important task of the Nobel Foundation is to market the prizes internationally and to oversee informal administration related to the prizes. The Foundation is not involved in the process of selecting the Nobel laureates. In many ways the Nobel Foundation is similar to an investment company, in that it invests Nobel's money to create a solid funding base for the prize and the administrative activities. The Nobel Foundation is exempt from all taxes in Sweden (since 1946) and from investment taxes in the United States (since 1953). Since 1980s, the Foundation's investments have become more profitable and as of 31 December 2007, the assets controlled by the Nobel Foundation amounted to 3.628 billion Swedish *kronor* (c. US $560 million).

According to the statutes, the Foundation should consist of a board of five Swedish or Norwegian citizens, with its seat in Stockholm. The Chairman of the Board should be appointed by the King in Council, with the other four members appointed by the trustees of the prize-awarding institutions. An Executive Director is chosen from among the board members, a Deputy Director is appointed by the King in Council, and two deputies are appointed by the trustees. However, since 1995 all the members of the board have been chosen by the trustees, and the Executive Director and the Deputy Director appointed by the board itself. As well as the board, the Nobel Foundation is made up of the prize-awarding institutions, the Royal Swedish Academy of Sciences, the Nobel Assembly, the Swedish Academy, and the Norwegian Nobel Committee, the trustees of these institutions, and auditors.

Wilhelm Röntgen

Wilhelm Conrad Röntgen was a German physicist, who, on 8 November 1895, produced and detected electromagnetic radiation in a wavelength range today known as X-rays or Röntgen rays, an achievement that earned him the first Nobel Prize in Physics in 1901.

Born	27 March 1845, Lennep, Prussia
Died	10 February 1923 (age 77), Munich, Germany
Nationality	German
Fields	Physics
Institutions	University of Strassburg, Hohenheim, University of Giessen, University of Würzburg, University of Munich
Alma mater	ETH Zurich, University of Zürich
Doctoral advisor	August Kundt
Doctoral students	Herman March, Abram Ioffe
Known for	X-rays
Notable awards	Nobel Prize in Physics (1901)

DISCOVERY OF X-RAYS (1901)

During 1895 Röntgen was investigating the external effects from the various types of vacuum tube equipment/ apparatus from Heinrich Hertz, Johann Hittorf, William Crookes, Nikola Tesla and Philipp von Lenard, when an electrical discharge is passed through them. In early November he was repeating an experiment with one of Lenard's tubes in which a thin aluminum window had been added to permit the cathode rays to exit the tube but a cardboard covering was added to protect the aluminum from damage by the strong electrostatic field that is necessary to produce the cathode rays. He knew the cardboard covering prevented light from escaping, yet Röntgen observed that the invisible cathode rays caused a fluorescent effect on a small cardboard screen painted with barium platinocyanide when it was placed close to the aluminum window. It occurred to Röntgen that the Hittorf-Crookes tube, which had a much thicker glass wall than the Lenard tube, might also cause this fluorescent effect.

In the late afternoon of 8 November 1895, Röntgen determined to test his idea. He carefully constructed a black cardboard covering similar to the one he had used on the Lenard tube. He covered the Hittorf-Crookes tube with the cardboard and attached electrodes to a Ruhmkorff coil to generate an electrostatic charge. Before setting up the barium platinocyanide screen to test his idea, Röntgen darkened the room to test the opacity of his cardboard cover. As he passed the Ruhmkorff coil charge through the tube, he determined that the cover was light-tight and turned to prepare the next step of the experiment. It was at this point that Röntgen noticed a faint shimmering from a bench a meter away from the tube. To be sure, he tried several more discharges and saw the same shimmering each time. Striking a match, he discovered the shimmering had come from the location of the barium platinocyanide screen he had been intending to use next.

Röntgen speculated that a new kind of ray might be responsible. 8 November was a Friday, so he took advantage of the weekend to repeat his experiments and make his first notes. In the following weeks he ate and slept in his laboratory as he investigated many properties of the new rays, he temporarily termed X-rays, using the mathematical designation for something unknown. Although the new rays would eventually come to bear his name in many languages where they became known as Röntgen Rays, he always preferred the term X-rays. Nearly two weeks after his discovery, he took the very first picture using X-rays of his wife's hand, Anna Bertha. When she saw her skeleton she exclaimed I have seen my death.

Hendrik Lorentz

 Hendrik Antoon Lorentz was a Dutch physicist who shared the 1902 Nobel Prize in Physics with Pieter Zeeman for the discovery and theoretical explanation of the Zeeman effect. He also derived the transformation equations subsequently used by Albert Einstein to describe space and time.

Born	18 July 1853, Arnhem, Netherlands
Died	4 February 1928 (age 74), Haarlem, Netherlands
Nationality	Netherlands
Fields	Physics
Alma mater	University of Leiden
Doctoral Advisor	Pieter Rijke
Doctoral students	Geertruida L. de Haas-Lorentz, Adriaan Fokker, Leonard Ornstein
Known for	Theory of electromagnetic radiation, Lorentz force
Notable awards	Nobel Prize in Physics (1902)

ELECTRODYNAMICS AND RELATIVITY (1902)

In 1895, with the attempt to explain the Michelson-Morley experiment, Lorentz proposed that moving bodies contract in the direction of motion. Lorentz worked on describing electromagnetic phenomena (the propagation of light) in reference frames that moved relative to each other. He discovered that the transition from one to another reference frame could be simplified by using a new time variable which he called *local time*. The local time depended on the universal time and the location under consideration. Lorentz's publications (of 1895 and 1899) made use of the term local time without giving a detailed interpretation of its physical relevance. In 1900, Henri Poincare called Lorentz's local time a "wonderful invention" and illustrated it by showing that clocks in moving frames are synchronized by exchanging light signals that are assumed to travel at the same speed against and with the motion of the frame.

In 1899, and again in his paper Electromagnetic phenomena in a system moving with any velocity smaller than that of light (1904), Lorentz added time dilation to his transformations and published what Poincare in 1905 named Lorentz transformations. It was apparently unknown to Lorentz that Joseph Larmor had used identical transformations to describe orbiting electrons in 1897. Larmor's and Lorentz's equations look unfamiliar, but they are algebraically equivalent to those presented by Poincare and Einstein in 1905. Lorentz's 1904 paper includes the covariant formulation of electrodynamics, in which electrodynamics phenomena in different reference frames are described by identical equations with well-defined transformation properties. The paper clearly recognizes the significance of this formulation, namely that the outcomes of electrodynamics experiments do not depend on the relative motion of the reference frame. The 1904 paper includes a detailed discussion of the increase of the inertial mass of rapidly moving objects. In 1905, Einstein would use many of the concepts, mathematical tools and results discussed to write his paper entitled Electrodynamics known today as the theory of special relativity. Because Lorentz laid the fundamentals for the work by Einstein, this theory was called the *Lorentz-Einstein theory* originally.

The increase of mass was the first prediction of special relativity to be tested, but from early experiments by Kaufmann it appeared that his prediction was wrong; this led Lorentz to the famous remark that he was at the end of his Latin. The confirmation of his prediction had to wait until 1908. In 1909, Lorentz published Theory of Electrons based on a series of lectures in Mathematical Physics he gave at Columbia University.

Pieter Zeeman

Pieter Zeeman, was a Dutch physicist who shared the 1902 Nobel Prize in Physics with Hendrik Lorentz for his.discovery of the Zeeman effect.

Born	25 May 1865, Zonnemaire, Netherlands
Died	9 October 1943 (age 78), Amsterdam, Netherlands
Nationality	Netherlands
Fields	Physics
Alma mater	University of Leiden Heike
Doctoral advisor	Kamerlingh Onnes
Known for	Zeeman effect
Notable awards	Nobel Prize in Physics (1902)

ZEEMAN EFFECT (1902)

The Zeeman effect is the splitting of a spectral line into several components in the presence of a static magnetic field. It is analogous to the Stark effect, the splitting of a spectral line into several components in the presence of an electric field. The Zeeman effect is very important applications such as nuclear magnetic resonance spectroscopy, electron spin resonance spectroscopy, magnetic resonance imaging (MRI) and Mossbauer spectroscopy. It may also be utilized to improve accuracy in Atomic absorption spectroscopy. When the spectral lines are absorption lines, the effect is called Inverse Zeeman effect. The Zeeman effect is named after the Dutch physicist Pieter Zeeman.

In most atoms, there exist several electron configurations with the same energy, so that transitions between these configurations and another correspond to a single spectral line.

The presence of a magnetic field breaks this degeneracy, since the magnetic field interacts differently with electrons with different quantum numbers, slightly modifying their energies. The result is that, where there were several configurations with the same energy, they now have different energies, giving rise to several very close spectral lines.

Without a magnetic field, configurations a, b and c have the same energy, as do d, e and f. The presence of a magnetic field (B) splits the energy levels. Therefore, a line produced by a transition from a, b or c and d, e or f will now split into several components between different combinations of a, b, c and d, e, f. However, not all transitions will be possible (in the dipole approximation), as governed by the selection rules. Since the distance between the Zeeman sub-levels is proportional to the magnetic field, this effect can be used by astronomers to measure the magnetic field of the Sun and other stars.

There is also an anomalous Zeeman effect that appears on transitions where the net spin of the electrons is not 0, the number of Zeeman sub-levels being even instead of odd if there's an uneven number of electrons involved. It was called "anomalous" because the electron spin had not yet been discovered, and so there was no satisfactory explanation for it at the time that Zeeman observed the effect. At higher magnetic fields the effect ceases to be linear. At even higher field strength, when the strength of the external field is comparable to the strength of the atom's internal field, electron coupling is disturbed and the spectral lines rearrange. This is called the Paschen-Back effect.

Henri Becquerel

Antoine Henri Becquerel was a French physicist, Nobel laureate, and the discoverer of radioactivity, for which he won the 1903 Nobel Prize in Physics.

Born	15 December 1852, Paris, France
Died	25 August 1908 (age 55), Le Croisic, Brittany, France
Nationality	French
Fields	Physics, chemistry
Institutions	Conservatoire des Arts et Metiers, École Polytechnique Muséum, National d'Histoire Naturelle
Alma mater	École Polytechnique, École des Ponts et Chaussées
Doctoral students	Marie Curie
Known for	Radioactivity
Notable awards	Nobel Prize in Physics (1903)
He is the father of Jean Becquerel, the son of A. E. Becquerel, and the grandson of Antoine César Becquerel.	

RADIOACTIVE DECAY (1903)

Radioactive decay is the process by which an unstable atomic nucleus loses energy by emitting ionizing particles or radiation. The emission is spontaneous in that the nucleus decays without collision with another particle. This decay, or loss of energy, results in an atom of one type, called the parent nuclide, transforming to an atom of a different type, named the daughter nuclide. For example: a carbon-14 atom (the parent) emits radiation and transforms to a nitrogen-14 atom (the daughter). This is a stochastic process on the atomic level, in that according to quantum mechanics it is impossible to predict when a given atom will decay. However given a large number of similar atoms the decay rate, on average, is predictable.

The SI unit of activity is the Becquerel (Bq). One Bq is defined as one transformation (or decay) per second. Since any reasonably-sized sample of radioactive material contains many atoms, a Bq is a tiny measure of activity; amounts on the order of GBq (gigabecquerel, 1×10^9 decays per second) or TBq (terabecquerel, 1×10^{12} decays per second) are commonly used. Another unit radioactivity is the curie, Ci, which was originally defined as the amount of radium emanation (^{222}Rn) in equilibrium with one gram of pure radium, isotope ^{226}Ra. At present it is equal, by definition, to the activity of any radionuclide decaying with a disintegration rate of 3.7×10^{10} Bq.

Radioactivity was first discovered in 1896 by the French scientist Henri Becquerel, while working on phosphorescent materials. These materials glow in the dark after exposure to light, and he thought that the glow produced in cathode ray tubes by X-rays might be connected with phosphorescence. He wrapped a photographic plate in black paper and placed various phosphorescent salts on it. All results were negative until he used uranium salts. The result with these compounds was a deep blackening of the plate. These radiations were called Becquerel Rays.

At first it seemed that the new radiation was similar to the then recently discovered X-rays. Further research by Becquerel, Marie Curie, Pierre Curie, Ernest Rutherford and others discovered that radioactivity was significantly more complicated. Different types of decay can occur, but Rutherford was the first to realize that they all occur with the same mathematical approximately exponential formula.

The early researchers also discovered that many other chemical elements besides uranium have radioactive isotopes. A systematic search for the total radioactivity in uranium ores also guided Marie Curie to isolate a new element polonium and to separate a new element radium from barium. The two elements' chemical similarity would otherwise have made them difficult to distinguish.

Pierre Curie

Pierre Curie was a French physicist, a pioneer in crystallography, magnetism, piezoelectricity and radioactivity, and Nobel laureate. In 1903 he received the Nobel Prize in Physics with his wife, Maria Skłodowska-Curie, and Henri Becquerel, in recognition of the extraordinary services they have rendered by their joint researches on the radiation phenomena discovered by Professor Henri Becquerel.

Born	15 May 1859, Paris, France
Died	19 April 1906 (age 46), Paris, France
Nationality	French
Fields	Physics
Alma mater	Sorbonne
Doctoral students	Paul Langevin, André-Louis Debierne, Marguerite Catherine Perey
Known for	Radioactivity
Notable awards	Nobel Prize in Physics (1903)
Married to Marie Curie (1895), their children include Irène Joliot-Curie and Ève Curie	

RADIOACTIVITY (1903)

The neutrons and protons that constitute nuclei, as well as other particles that may approach them, are governed by several interactions. The strong nuclear force, not observed at the familiar macroscopic scale, is the most powerful force over subatomic distances. The electrostatic force is almost always significant, and in the case of beta decay, the weak nuclear force is also involved.

The interplay of these forces produces number of different phenomena in which energy may be released by rearrangement of particles. Some configurations of the particles in a nucleus have the property that, should they shift ever so slightly, the particles could rearrange into a lower-energy arrangement and release some energy. One might draw an analogy with a snowfield on a mountain: while friction between the ice crystals may be supporting the snow's weight, the system is inherently unstable with regard to a state of lower potential energy. A disturbance would thus facilitate the path to a state of greater entropy: the system will move towards the ground state, producing heat, and the total energy will be distributable over a larger number of quantum states.

In the case of an excited atomic nucleus, the arbitrarily small disturbance comes from quantum vacuum fluctuations. A radioactive nucleus is unstable, and can thus *spontaneously* stabilize to a less-excited system. The resulting transformation alters the structure of the nucleus and results in the emission of either a photon or a high-velocity particle which has mass.

The dangers of radioactivity and of radiation were not immediately recognized. Acute effects of radiation were first observed in the use of X-rays when electrical engineer and physicist Nikola Tesla intentionally subjected his fingers to X-rays in 1896. He published his observations concerning the burns that developed, though he attributed them to ozone rather than to X-rays. His injuries healed later.

The genetic effects of radiation, including the effects on cancer risk, were recognized much later.

Marie Curie

Marie Skłodowska Curie was a physicist and chemist of Polish, upbringing and subsequent French citizenship. She was a pioneer in the field of radioactivity and the first person honored with two Nobel Prizes in physics and chemistry. She was also the first female professor at the University of Paris.

Born	7 November 1867, Warsaw, Vistula Land, Russian Empire
Died	4 July 1934 (age 66), Passy, France
Nationality	Polish
Citizenship	Russian, later French
Fields	Physics, chemistry
Institutions	University of Paris
Alma mater	University of Paris, ESPCI
Doctoral advisor	Henri Becquerel
Doctoral students	André-Louis Debierne, Óscar Moreno, Marguerite Catherine Perey
Known for	Radioactivity, polonium, radium
Notable awards	Nobel Prize in Physics (1903) Davy Medal (1903) Matteucci Medal (1904) Nobel Prize in Chemistry (1911)

She is the only person to win Nobel Prizes in two sciences. She was the wife of Pierre Curie, and the mother of Irene Joliot-Curie and Ève Curie

POLONIUM (1903)

Polonium is a chemical element with the symbol Po and atomic number 84, discovered in 1898 by Marie Skłodowska-Curie and Pierre Curie. A rare and highly radioactive metalloid, polonium is chemically similar to bismuth and tellurium, and it occurs in uranium ores. Polonium has been studied for possible use in heating spacecraft. It is unstable, all isotopes of polonium are radioactive.

Polonium has 33 known isotopes, all of which are radioactive. They have atomic masses that range from 188 to 220. ^{210}Po (half-life 138.376 days) is the most widely available. ^{209}Po (half-life 103 years) and ^{208}Po (half-life 2.9 years) can be made through the alpha, proton, or deuteron bombardment of lead or bismuth in a cyclotron.

RADIUM

Radium is a radioactive chemical element which has the symbol Ra and atomic number 88. Its appearance is almost pure white, but it readily oxidizes on exposure to air, turning black. Radium is an alkaline earth metal that is found in trace amounts in uranium ores. Its most stable isotope, ^{226}Ra, has a half-life of 1602 years and decays into radon gas.

The heaviest of the alkaline earth metals, radium is intensely radioactive and resembles barium in its chemical behavior. This metal is found in tiny quantities in the uranium ore pitch blende, and various other uranium minerals. Radium preparations are remarkable for maintaining themselves at a higher temperature than their surroundings, and for their radiations, which are of three kinds: alpha particles, beta particles, and gamma rays.

When freshly prepared, pure radium metal is brilliant white, but blackens when exposed to air (probably due to nitride formation). Radium is luminescent (giving a faint blue colour), reacts violently with water and oil to form radium hydroxide and is slightly more volatile than barium. The normal phase of radium is a solid.

Radium (Latin *radius*, ray) was discovered by Marie Skłodowska-Curie and her husband Pierre in 1898 in pitch blende coming from North Bohemia, in the Czech Republic (area around Jáchymov).

In 1910, radium was isolated as a pure metal by Curie and André-Louis Debierne through the electrolysis of a pure radium chloride solution by using a mercury cathode and distilling in an atmosphere of hydrogen gas.

John Strutt, 3rd Baron Rayleigh

 John William Strutt, 3rd Baron Rayleigh, was an English physicist who, with William Ramsay, discovered the element argon, an achievement for which he earned the Nobel Prize for Physics in 1904.

He also discovered the phenomenon now called Rayleigh scattering, explaining why the sky is blue, and predicted the existence of the surface waves now known as Rayleigh waves. In 1910 Lord Rayleigh discovered that an electrical discharge in nitrogen gas produced active nitrogen, an allotrope considered to be monatomic. The whirling cloud of brilliant yellow light produced by his apparatus reacted with quicksilver to produce explosive mercury nitride.

Born	12 November 1842, Langford Grove, Maldon, Essex, England
Died	30 June 1919 (age 76), Terling Place, Witham, Essex, England
Nationality	United Kingdom
Fields	Physics
Institutions	University of Cambridge
Alma mater	University of Cambridge
Doctoral advisor	Edward John Routh
Doctoral students	J. J. Thomson, George Paget Thomson, Jagdish Chandra Bose
Known for	Discovery of argon, Rayleigh waves, Rayleigh scattering, Rayleigh criterion, Duplex Theory, Theory of Sound, Rayleigh flow
Notable awards	Nobel Prize in Physics (1904)

DISCOVERY OF ARGON (1904)

Argon is a chemical element represented by the symbol Ar, has atomic number 18 and is the third element in group 18 of the periodic table. It is the third most common gas in the Earth's atmosphere, at 0.93%, making it more common than carbon dioxide. The complete octet in the outer atomic shell makes it stable and resistant to bonding with other elements. Its triple point temperature of 83.8058 K is a defining fixed point in the International Temperature Scale of 1990.

Argon constitutes 0.934% by volume and 1.29% by mass of the Earth's atmosphere, and air is the primary raw material used by industry to produce purified argon products. Argon is isolated from air by fractionation, most commonly by cryogenic fractional distillation, a process that also produces purified nitrogen, oxygen, neon, krypton and xenon.

Argon (Greek meaning inactive) was suspected to be present in air by Henry Cavendish in 1785 but was not isolated until 1894 by Lord Rayleigh and Sir William Ramsay in Scotland in an experiment in which they removed all of the oxygen, carbon dioxide, water and nitrogen from a sample of clean air. They had determined that nitrogen produced from chemical compounds was one-half percent lighter than nitrogen from the atmosphere. The difference seemed insignificant, but it was important enough to attract their attention for many months. They concluded that there was another gas in the air mixed in with the nitrogen. Argon was also encountered in 1882 through independent research of H. F. Newall and W.N. Hartley. Each observed new lines in the colour spectrum of air but were unable to identify the element responsible for the lines. Argon became the first member of the noble gases to be discovered. The symbol for argon is now Ar, but up until 1957 it was A.

Argon has approximately the same solubility in water as oxygen gas and is 2.5 times more soluble in water than nitrogen gas. Argon is colourless, odorless, and nontoxic as a solid, liquid, and gas. Argon is inert under most conditions and forms no confirmed stable compounds at room temperature.

Philipp Lenard

Philipp Eduard Anton von Lenard, known in Hungarian as Lénárd Fülöp Eduárd Antal, was a Hungarian-German physicist and the winner of the Nobel Prize for Physics in 1905 for his research on cathode rays and the discovery of many of their properties. He was also an active proponent of Nazi ideology.

Born	7 June 1862, Pozsony, Kingdom of Hungary, Austrian Empire
Died	20 May 1947 (age 84), Messelhausen, Germany
Citizenship	Hungarian, in Austria-Hungary (1862–1907), German (1907–1947)
Nationality	Hungarian, German
Fields	Physics
Institutions	University of Budapest, University of Breslau, University of Aachen, University of Heidelberg, University of Kiel
Alma mater	University of Heidelberg
Doctoral advisor	Robert Bunsen
Known for	Cathode rays
Notable awards	Nobel Prize in Physics (1905)

CATHODE RAY (1905)

In 1838, Michael Faraday passed a current through a rarefied air filled glass tube and noticed a strange light arc with its beginning at the cathode (negative electrode) and its end almost at the anode (positive electrode). In 1857, German physicist and glassblower Heinrich Geissler sucked even more air out with an improved pump, to a pressure of around 10^{-3} atm and found that, instead of an arc, the glow filled the tube. The voltage applied between the two electrodes of the tubes, generated by an induction coil, was anywhere between a few kilovolts and 100 kV. These were called Geissler tubes, similar to today's neon lights. The explanation of these effects was that the high voltage accelerated electrically charged atoms (ions) were naturally present in the air of the tube. At low pressure, there was enough space between the gas atoms that the ions could accelerate to high enough speeds that when they struck another atom they knocked electrons off it, creating more positive ions and free electrons in a chain reaction. The positive ions were all attracted to the cathode. When they struck it, they knocked many electrons out of the metal. The free electrons were all attracted to the anode.

In the Geissler tubes, there was so much air that the electrons could only travel a tiny distance before colliding with an atom. The electrons in these tubes moved in a slow diffusion process, never gaining much speed, so these tubes did not produce cathode rays. The glow in the gas was caused when the electrons or ions struck gas atoms, exciting their orbital electrons to higher energy levels. The electrons released this energy as light. This process is called fluorescence.

By the 1870s, British physicist William Crookes and others were able to evacuate tubes to a lower pressure, below 10^{-6} atm. These were called Crookes tubes. Faraday had been the first to notice a dark space just in front of the cathode, where there was no luminescence. This came to be called the cathode dark space, Faraday dark space or Crookes dark space. Crookes found that as he pumped more air out of the tubes, the Faraday dark space spread down the tube from the cathode toward the anode, until the tube was totally dark. But at the anode (positive) end of the tube, the glass of the tube itself began to glow.

Cathode rays themselves are invisible, but this accidental fluorescence allowed researchers to notice that objects in the tube in front of the cathode, such as the anode, cast sharp-edged shadows on the glowing back wall. In 1869, German physicist Johann Hittorf was first to realize that something must be travelling in straight lines from the cathode to cast the shadows. Eugen Goldstein named them *cathode rays*.

J. J. Thomson

 Sir Joseph John (J. J. Thomson) was a British physicist and Nobel laureate, credited for the discovery of the electron and of isotopes, and the invention of the mass spectrometer. He was awarded the 1906 Nobel Prize in Physics for the discovery of the electron and his work on the conduction of electricity in gases.

Born	18 December 1856, Cheetham Hill, Manchester, UK
Died	30 August 1940 (age 83), Cambridge, UK
Nationality	British
Fields	Physics
Institutions	Cambridge University
Alma mater	University of Manchester, University of Cambridge
Academic advisors	John Strutt (Rayleigh), Edward John Routh
Notable students	Charles Glover Barkla, Charles T. R. Wilson, Ernest Rutherford, Francis William Aston, John Townsend, J. Robert Oppenheimer, Owen Richardson, William Henry Bragg, H. Stanley Allen, John Zeleny, Daniel Frost Comstock, Max Born, T. H. Laby, Paul Langevin, Balthasar van der Pol, Geoffrey Ingram Taylor
Known for	Plum pudding model, Discovery of electron, discovery of isotopes, mass spectrometer invention, first m/e measurement, proposed first waveguide, Thomson scattering, Thomson problem, Coining term 'delta ray', Coining term 'epsilon radiation' Thomson (unit)
Notable awards	Nobel Prize in Physics (1906)

ELECTRON (1906)

The electron is a sub-atomic particle carrying a negative electric charge. It has no known components or sub-structure, and therefore is believed to be an elementary particle. An electron has a mass that is approximately 1/1836 that of the proton. The intrinsic angular momentum (spin) of the electron is a half integer value in units of ℏ, which means that it is a fermion. The antiparticle of the electron is called the positron, which is identical to the electron except that it carries electrical and other charges of the opposite sign. When an electron collides with a positron, they may either scatter off each other or be totally annihilated, producing a pair (or more) of gamma ray photons. Electrons, which belong to the first generation of the lepton particle family, participate in gravitational, electromagnetic and weak interactions. Electrons, like all matter, have quantum mechanical properties of both a particle and a wave, so they can collide with other particles and be diffracted like light. However, this duality is best demonstrated in experiments with electrons, due to their tiny mass. Since an electron is a fermion, no two electrons can occupy the same quantum state, in accordance with the Pauli exclusion principle.

The concept of an indivisible amount of electric charge was theorized to explain the chemical properties of atoms, beginning in 1838 by British natural philosopher Richard Laming; the name *electron* was introduced for this charge in 1894 by Irish physicist George Johnstone Stoney. The electron was identified as a particle in 1897 by J. J. Thomson and his team of British physicists.

In many physical phenomena, such as electricity, magnetism, and thermal conductivity, electrons play an essential role. An electron in motion relative to an observer generates a magnetic field, and will be deflected by external magnetic fields. When an electron is accelerated, it can absorb or radiate energy in the form of photons. Electrons, together with atomic nuclei made of protons and neutrons, make up atoms. However, electrons contribute less than 0.06% to an atom's total mass. The attractive Coulomb force between an electron and a proton causes electrons to be bound into atoms. The exchange or sharing of the electrons between two or more atoms is the main cause of chemical bonding.

Albert Abraham Michelson

 Albert Abraham Michelson was an American physicist known for his work on the measurement of the speed of light and especially for the Michelson-Morley experiment. In 1907 he received the Nobel Prize in Physics. He became the first American to receive the Nobel Prize in sciences.

Born	19 December 1852, Strzelno, Kingdom of Prussia
Died	9 May 1931 (age 78), Pasadena, California
Nationality	United States
Fields	Physics
Institutions	Case Western Reserve University, Clark University, University of Chicago
Alma mater	United States Naval Academy, University of Berlin
Doctoral advisor	Hermann Helmholtz
Doctoral students	Robert Millikan
Known for	Speed of light, Michelson-Morley experiment
Notable awards	Nobel Prize in Physics (1907)

MICHELSON-MORLEY EXPERIMENT (1907)

The Michelson-Morley experiment was performed in 1887 by Albert Michelson and Edward Morley at what is now Case Western Reserve University. Its results are generally considered to be the first strong evidence against the theory of a luminiferous aether. The experiment has also been referred to as the kicking-off point for the theoretical aspects of the Second Scientific Revolution.

Michelson had a solution to the problem of how to construct a device sufficiently accurate to detect aether flow. The device he designed, later known as an interferometer, sent a single source of white light through a half-silvered mirror that was used to split it into two beams traveling at right angles to one another. After leaving the splitter, the beams travelled out to the ends of long arms where they were reflected back into the middle on small mirrors. They then recombined on the far side of the splitter in an eyepiece, producing a pattern of constructive and destructive interference based on the spent time to transit the arms. If the Earth is traveling through an ether medium, a beam reflecting back and forth parallel to the flow of ether would take longer than a beam reflecting perpendicular to the ether because the time gained from traveling downwind is less than that lost traveling upwind. The result would be a delay in one of the light beams that could be detected when the beams were recombined through interference. Any slight change in the spent time would then be observed as a shift in the positions of the interference fringes. If the aether were stationary relative to the Sun, then the Earth's motion would produce a fringe shift of 4% the size of a single fringe.

In 1881, while he was still in Germany, Michelson had used an experimental device to make several measurements, in which he noticed that the expected shift of 0.04 was not seen, and a smaller shift of (at most) about 0.02 was seen. However his apparatus was a prototype, and had experimental errors far too large to say anything about the aether wind. For a measurement of the aether wind, a much more accurate and tightly controlled experiment would have to be carried out. The prototype was, however, successful in demonstrating that the basic method was feasible.

In 1887 he then combined efforts with Edward Morley and spent a considerable amount of time and money creating an improved version with more than enough accuracy to detect the drift. In their experiment, the light was repeatedly reflected back and forth along the arms, increasing the path length to 11 m. At this length, the drift would be about 0.4 fringes. To make that easily detectable, the apparatus was located in a closed room in the basement of a stone building, eliminating most thermal and vibrational effects. Vibrations were further reduced by building the apparatus on top of a huge block of marble, which was then floated in a pool of mercury. They calculated that effects of about 1/100th of a fringe would be detectable.

Gabriel Lippmann

 Jonas Ferdinand Gabriel Lippmann was a Franco-Luxembourgish physicist and inventor, and Nobel laureate in physics for his method of reproducing colours photographically based on the phenomenon of interference, later known as the Lippmann plate.

Born	16 August 1845, Bonnevoie, Luxembourg
Died	13 July 1921 (age 75) SS France, Atlantic Ocean
Nationality	France
Fields	Physics
Institutions	Sorbonne
Alma mater	École Normale
Known for	Color photography, Lippmann plate, Lippmann electrometer
Notable awards	Nobel Prize in Physics (1908)

COLOUR PHOTOGRAPHY (1908)

Colour photography is photography that uses media capable of representing colours, which are produced chemically during the photographic processing phase. It is contrasted with black and white photography, which uses media capable only of showing shades of gray. It does not include hand coloured or Photochrome photographs either. Some examples of colour photography include prints, colour negatives, transparencies and slides, and roll and sheet films.

The Lippmann plate was an early form of colour photography developed in 1891 by Gabriel Lippmann, a physicist. Lippmann won the Nobel Prize in Physics in 1908 for its development.

A glass plate is coated with transparent and grain less silver emulsion. It is the uncoated side which is exposed to the light with the emulsion in contact with a reflecting surface such as mercury. The incident light is reflected back on itself causing interference. This establishes standing waves in the emulsion at half the wavelength of the incident light which react with the photosensitive emulsion. The plate is then processed so that the recording changes in the index of refraction of the gelatin. These changes of index of refraction reflect the light by a process called Bragg diffraction.

The colour image can only be viewed in the reflection of a diffuse light source from the plate, making the field of view limited, and it cannot be copied. The technique was very insensitive with the emulsions of the time and it never came into general use.

Lippmann photographic techniques are being developed to produce images which can easily be viewed, but not copied, for security purposes.

One of Lippmann's early discoveries was the relationship between electrical and capillary phenomena which allowed him to develop a sensitive capillary electrometer, subsequently known as the Lippmann electrometer which was used in the first ECG machine. In a paper delivered to the Philosophical Society of Glasgow on 17 January 1883, John G. M'Kendrick described the apparatus as follows: Lippmann's electrometer consists of a tube of ordinary glass, 1 m long and 7 mm in diameter, open at both ends, and kept in the vertical position by a stout support. The lower end is drawn into a capillary point, until the diameter of the capillary is 0.005 of a mm. The tube is filled with mercury, and the capillary point is immersed in dilute sulphuric acid, and in the bottom of the vessel containing the acid there is a little more mercury. A platinum wire is put into connection with the mercury in each tube, and, finally, arrangements are made by which the capillary point can be seen with a microscope magnifying 250 diameters. Such an instrument is very sensitive; and Lippmann states that it is possible to determine a difference of potential so small as that of one 10,080th of a Daniell. It is thus a very delicate means of observing and of measuring minute electromotive forces.

Guglielmo Marconi

Guglielmo Marconi was an Italian inventor, best known for his development of a radio telegraph system, which served as the foundation for the establishment of numerous affiliated companies worldwide. He shared the 1909 Nobel Prize in Physics with Karl Ferdinand Braun, in recognition of their contributions to the development of wireless telegraphy and was ennobled in 1924 as Marchese Marconi.

Born	25 April 1874, Palazzo Marescalchi, Bologna, Italy
Died	20 July 1937 (age 63), Rome, Italy
Nationality	Italian
Fields	Physical and electrical science
Known for	Radio
Notable awards	Nobel Prize in Physics (1909)

RADIO (1909)

Radio is the transmission of signals by modulation of electromagnetic waves with frequencies below those of visible light. Electromagnetic radiation travels by means of oscillating electromagnetic fields that pass through the air and the vacuum of space. Information is carried by systematically changing some property of the radiated waves, such as amplitude, frequency, phase, or pulse width. When radio waves pass on electrical conductor, the oscillating fields induce an alternating current in the conductor. This can be detected and transformed into sound or other signals that carry information.

The meaning and usage of the word radio has developed in parallel with the developments within the field and can be seen to have three distinct phases: electromagnetic waves and experimentation; wireless communication and technical development; and radio broadcasting and commercialization. Many individuals, inventors, engineers, developers, businessmen contributed to produce the modern idea of radio and thus the origins and invention are multiple and controversial. Early radio could not transmit sound or speech and was called the wireless telegraph.

Development from a laboratory demonstration to a commercial entity spanned several decades and required the efforts of many practitioners. In 1878, David E. Hughes noticed that sparks could be heard in a telephone receiver when experimenting with his carbon microphone. He developed this carbon-based detector further and eventually could detect signals over a few hundred yards. He demonstrated his discovery to the Royal Society in 1880, but was told it was merely induction, and therefore abandoned further research. Experiments, later patented, were undertaken by Thomas Edison and his employees of Menlo Park. Edison applied in 1885 to the U.S. Patent Office for his patent on an electrostatic coupling system between elevated terminals. The patent was granted as U.S. Patent 465,971 on 29 December 1891. The Marconi Company would later purchase rights to the Edison patent to protect them legally from lawsuits.

In 1893, in St. Louis, Missouri, Nikola Tesla made devices for his experiments with electricity. Addressing the *Franklin Institute* in Philadelphia and the *National Electric Light Association*, he described and demonstrated the principles of his wireless work. The descriptions contained all the elements that were later incorporated into radio systems before the development of the vacuum tube. He initially experimented with magnetic receivers, unlike the coherers used by Guglielmo Marconi and other early experimenters.

A demonstration of wireless telegraphy took place in the lecture theater of the Oxford University Museum of Natural History on 14 August 1894, carried out by Professor Oliver Lodge and Alexander Muirhead. During the demonstration a radio signal was sent from the neighboring Clarendon laboratory building, and received by apparatus in the lecture theater.

Karl Ferdinand Braun

Karl Ferdinand Braun was a German inventor, physicist and Nobel laureate in physics. Braun contributed significantly to the development of the radio and TV technology. he shared with Guglielmo Marconi the 1909 Nobel Prize in Physics.

Born	6 June 1850, Fulda, Hessen-Kassel, Germany
Died	20 April 1918 (age 67), Brooklyn, New York, USA
Nationality	German
Fields	Physics
Institutions	University of Karlsruhe, University of Marburg University of Strassburg, University of Tübingen University of Würzburg
Alma mater	University of Marburg, University of Berlin
Doctoral advisor	August Kundt
Doctoral students	Leonid Isaakovich Mandelshtam, Albert Schweizer
Known for	Cathode ray tube, Cat's whisker diode
Notable awards	Nobel Prize in Physics (1909)

A German physicist who shared the Nobel Prize for Physics in 1909 with Guglielmo Marconi for the development of wireless telegraphy. Braun received his doctorate from the University of Berlin in 1872. After appointments at Wurzburg, Leipzig, Marburg, Karlsruhe, and Tubingen, he became director of the Physical Institute and professor of physics at the University of Strasbourg in 1895. Braun was recognized by the Nobel committee for his improvement of Marconi's transmitting system. In early wireless transmission, the antenna was directly in the power circuit and broadcasting was limited to a range of about 15 km. Braun solved this problem by producing a spark less antenna circuit that linked transmitter power to the antenna circuit inductively. This invention greatly increased the broadcasting range of a transmitter and has been applied to radar, radio, and television. Braun's discovery of crystalline materials that act as rectifiers, allowing current to flow in one direction only led to the development of crystal radio receivers. He is also known as the developer of the cathode-ray oscilloscope. He demonstrated the first oscilloscope in 1897, after working on high-frequency alternating currents. Cathode-ray tubes had previously been characterized by uncontrolled rays; Braun succeeded in producing a narrow stream of electrons, guided by the means of alternating voltage that could trace patterns on a fluorescent screen. This invention, the forerunner of the television tube and radarscope, also became an important laboratory research instrument. He went to New York City in 1915 to testify in a radio-related patent case. He was detained there because of his German citizenship when the U.S. entered World War I in 1917; he died before the war ended.

CATHODE RAY TUBE (1909)

The Cathode Ray Tube (CRT) is a vacuum tube containing an electron gun (a source of electrons) and a fluorescent screen, with internal or external means to accelerate and deflect the electron beam, used to create images in the form of light emitted from the fluorescent screen. The image may represent electrical waveforms (oscilloscope), pictures (television, computer monitor), radar targets and others. The CRT uses an evacuated glass envelope which is large, deep, heavy, and relatively fragile.

CAT'S-WHISKER DETECTOR

A cat's whisker detector is an early electronic component consisting of a thin wire that lightly touches a crystal of semiconducting mineral to make a crude contact-junction rectifier. This device was used as the detector in early crystal radios, from about 1906 through the 1940s, and gave this type of radio receiver its name. It was the first type of semiconductor diode, and in fact the first semiconductor electronic device. The term cat's whisker was also sometimes used to describe the crystal receiver itself. Cat's whisker detectors are obsolete and are now only used in antique or antique-reproduction radios.

Johannes Diderik van der Waals

Johannes Diderik van der Waals was a Dutch physicist and thermodynamicist famous for his work on an equation of state for gases and liquids. Johannes Diderik van der Waals awarded Nobel Prize in Physics in 1910.

Born	23 November 1837, Leiden, Netherlands
Died	8 March 1923 (age 85), Amsterdam, Netherlands
Nationality	Netherlands
Fields	Physics
Institutions	University of Amsterdam
Alma mater	University of Leiden
Doctoral advisor	Pieter Rijke
Doctoral students	Diederik Korteweg
Known for	Equation of state, intermolecular forces
Notable awards	Nobel Prize in Physics (1910)

VAN DER WAALS EQUATION (1910)

The van der Waals equation is an equation of state for a fluid composed of particles that have a non-zero size and a pair wise attractive inter-particle force (such as the van der Waals force). It was derived by Johannes Diderik van der Waals in 1873, who received the Nobel Prize in 1910 for his work on the equation of state for gases and liquids. The equation is based on a modification of the ideal gas law and approximates the behavior of real fluids, taking into account the nonzero size of molecules and the attraction between them.

The first form of this equation is

$$\left(p + \frac{a'}{v^2}\right)(v - b') = kT$$

Where, p is the pressure of the fluid, v is the volume of the container holding the particles divided by the total number of particles, k is Boltzmann's constant, T is the absolute temperature

a' is a measure for the attraction between the particles and b' is the average volume excluded from v by a particle.

Upon introduction of the Avogadro constant N_A, the number of moles n, and the total number of particles nN_A, the equation can be cast into the second (better known) form

$$\left(p + \frac{n^2 a}{V^2}\right)(V - nb) = nRT,$$

where, p is the pressure of the fluid, V is the total volume of the container containing the fluid

a is a measure of the attraction between the particles $a = N_A^2 a'b$, b is the volume excluded by a mole of particles $b = N_A^2 b'$, n is the number of moles, R is the universal gas constant, $R = N_A k$, T is the absolute temperature.

A careful distinction must be drawn between the volume available *to* a particle and the volume *of* a particle. In particular, in the first equation v refers to the empty space available per particle. That is, v is the volume V of the container divided by the total number nN_A of particles. The parameter b', on the other hand, is proportional to the proper volume of a single particle—the volume bounded by the atomic radius. This is the volume to be subtracted from v because of the space taken up by one particle. In van der Waals' original derivation, given below, b' is four times the proper volume of the particle. Observe further that the pressure p goes to infinity when the container is completely filled with particles so that there is no void space left for the particles to move. This occurs when $V = n\, b$.

Wilhelm Wien

 Wilhelm Carl Werner Otto Fritz Franz Wien, was a German physicist who, in 1893, used theories about heat and electromagnetism to deduce Wien's displacement law, which calculates the emission of a blackbody at any temperature from the emission at any one reference temperature.

He also formulated an expression for the black-body radiation which is correct in the photon-gas limit. His arguments were based on the notion of adiabatic invariance, and were instrumental for the formulation of quantum mechanics. Wien received the 1911 Nobel Prize for his work on heat radiation.

Born	13 January 1864, Fischhausen, East Prussia
Died	30 August 1928 (age 64), Munich, Germany
Nationality	German
Fields	Physics
Institutions	University of Giessen, University of Würzburg University of Munich, RWTH Aachen
Alma mater	University of Göttingen, University of Berlin
Doctoral advisor	Hermann von Helmholtz
Doctoral students	Karl Hartmann, Gabriel Holtsmark
Known for	Blackbody radiation, Wien's law
Notable awards	Nobel Prize in Physics (1911)

WIEN'S DISPLACEMENT LAW (1911)

Thermal radiation is electromagnetic radiation emitted from a material which is due to the heat of the material, the characteristics of which depend on its temperature. An example of thermal radiation is the infrared radiation emitted by a common household radiator or electric heater. A person near a raging bonfire will feel the radiated heat of the fire, even if the surrounding air is very cold. Thermal radiation is generated when heat from the movement of charges in the material is converted to electromagnetic radiation. Sunshine, or solar radiation, is thermal radiation from the extremely hot gases of the Sun, and this radiation heats the Earth. The Earth also emits thermal radiation, but at a much lower intensity because it is cooler. The balance between heating by incoming solar thermal radiation and cooling by the Earth's outgoing thermal radiation is the primary process that determines the Earth's overall temperature. If the object is a black-body in thermodynamic equilibrium, the radiation is termed as black-body radiation. The emitted wave frequency of the black-body thermal radiation is described by a probability distribution depending only on temperature, and for a genuine black-body in thermodynamic equilibrium is given by Planck's law of radiation. Wien's law gives the most likely frequency of the emitted radiation, and the Stefan-Boltzmann law gives the radiant intensity.

Wien's displacement law of physics was used to describe the spectrum of thermal radiation. This law was first derived by Wilhelm Wien in 1896. The equation does accurately describe the short wavelength spectrum of thermal emission from objects, but it fails to accurately fit the experimental data for long wavelengths emission. Wien's displacement law states that the wavelength distribution of radiated heat energy from a black-body at any temperature has essentially the same shape as the distribution at any other temperature, except that each wavelength is displaced, or moved over, on the graph. The average heat energy in each mode with frequency γ only depends on the combination γ/T. Restated in terms of the wavelength $\lambda = c/\gamma$, the distributions at corresponding wavelengths are related, where corresponding wavelengths are at locations proportional to $1/T$. From this general law, it follows that there is an inverse relationship between the wavelength of the peak of the emission of a black-body and its temperature when expressed as a function of wavelength, and this less powerful consequence is often also called Wien's displacement law. $\lambda_{max} = b/T$

where λ_{max} is the peak wavelength, T is the absolute temperature of the black-body, and b is a constant of proportionality called Wien's displacement constant, equal to $2.8977685(51) \times 10^{-3}$ m K The two digits between the parentheses denote the uncertainty in the two least significant digits of the mantissa.

Gustaf Dalén

Nils Gustaf Dalén, was a Swedish Nobel laureate and industrialist, the founder of the AGA company and inventor of the AGA cooker and the Dalén light. In 1912 he was awarded the Nobel Prize in Physics for his invention of automatic regulators for use in conjunction with gas accumulators for illuminating lighthouses and buoys.

Born	30 November 1869, Stenstorp, Västergötland, Sweden
Died	9 December 1937 (age 68), Lidingö, Stockholm, Sweden
Nationality	Swedish
Fields	Physics, mechanical engineering
Institutions	AGA
Alma mater	Chalmers University of Technology, Polytechnikum, Zürich
Known for	Sun valve and other lighthouse regulators
Notable awards	Nobel Prize in Physics (1912)

During a test in late 1912 to figure out the maximum pressure capacity for the acetylene accumulators the last tube in a series of tubes did not explode but showed a small flame coming out. As Dalén got closer to investigate the flame and the accumulator pressure it suddenly exploded and Dalén was heavily injured and blinded. It took several months before he returned to the company.

Dalén exploited the new fuel, developing the Dalén light which incorporated another invention, the 'Sun valve'. This device allowed the light to operate only at night, conserving fuel, and extending their service life to over a year.

The 'Dalén Flasher' was a device that, except for a small pilot light, only consumed gas during the flash stage. This reduced gas consumption by more than 90%. The AGA lighthouse equipment worked without any type of electric supply and was thus extremely reliable.

DALÉN LIGHT/SUN VALVE (1912)

The Dalén light was the predominant form of light source in lighthouses from the 1900s through the 1960s, when electric lighting had become dominant. The system was invented by Gustaf Dalén and marketed by his company AGA. Dalén later invented the AGA cooker in 1922. The Dalén light is notable because of its Sun valve, which earned its inventor the Nobel Prize in physics.

The Carbide lamp was developed in the early 1900s. While the lamps proved useful in many applications, the problem of safely storing acetylene meant they needed regular refilling which constrained their use in applications such as lighthouses.

A sun valve (*Solventil*, "solar valve") is a form of flow control valve, notable because it earned its inventor Gustaf Dalén the Nobel Prize in physics.

The valve formed part of the Dalén light which was used in lighthouses from the 1900s through to the 1960s by which time electric lighting came to dominate.

The valve is controlled by four metal rods enclosed in a glass tube. One rod is blackened, while the others are highly polished. As morning sunlight light fell onto all of the rods, the absorbed heat of the Sun allowed the unequally expanding dark rod to cut the gas supply.

Heike Kamerlingh Onnes

Heike Kamerlingh Onnes, was a Dutch physicist and Nobel laureate. He pioneered refrigeration techniques and explored how materials behaved when cooled to nearly absolute zero. This led to his discovery of superconductivity for certain materials, electrical resistance abruptly vanishes at very low temperatures.

Born	21 September 1853, Groningen, Netherlands
Died	21 February 1926 (age 72), Leiden, Netherlands
Nationality	Netherlands
Fields	Physics
Institutions	University of Leiden
Alma mater	Heidelberg University, University of Groningen
Doctoral advisor	Rudolf Adriaan Mees
Academic advisors	Robert Bunsen, Gustav Kirchhoff, Johannes Bosscha
Doctoral students	Jacob Clay, Claude Crommelin, Wander de Haas Johannes Kuenen, Remmelt Sissingh, Ewoud van Everdingen, Jules Verschaffelt, Pieter Zeeman
Known for	Onnes-effect, superconductivity
Notable awards	Nobel Prize in Physics (1913)

SUPERCONDUCTIVITY (1913)

Superconductivity is an electrical resistance of exactly zero which occurs in certain materials below a characteristic temperature. It was discovered by Heike Kamerlingh Onnes in 1911. Like ferromagnetism and atomic spectral lines, superconductivity is a quantum mechanical phenomenon. It is also characterized by a phenomenon called the Meissner effect, the ejection of any sufficiently weak magnetic field from the interior of the superconductor as it transitions into the superconducting state. The occurrence of the Meissner effect indicates that superconductivity cannot be understood simply as the idealization of perfect conductivity in classical physics.

The electrical resistivity of a metallic conductor decreases gradually as the temperature is lowered. However, in ordinary conductors such as copper and silver, this decrease is limited by impurities and other defects. Even near absolute zero, a real sample of copper shows some resistance. In a superconductor however, despite these imperfections, the resistance drops abruptly to zero when the material is cooled below its critical temperature. An electric current flowing in a loop of superconducting wire can persist indefinitely with no power source.

Superconductivity occurs in many materials: simple elements like tin and aluminum, various metallic alloys and some heavily-doped semiconductors. Superconductivity does not occur in noble metals like gold and silver, nor in pure samples of ferromagnetic metals.

In 1986, it was discovered that some cuprate-perovskite ceramic materials have critical temperatures above 90 kelvins (-183.15 degrees Celsius). These high-temperature superconductors renewed interest in the topic because of the prospects for improvement and potential room-temperature superconductivity. From a practical perspective, even 90 kelvins is relatively easy to reach with the readily available liquid nitrogen (boiling point 77 kelvins), resulting in more experiments and applications.

Onnes-Effect: A Rollin film, named after Bernard V. Rollin, is a 30 nm-thick liquid film of helium in the helium II state. It exhibits a creeping effect in response to surfaces extending past the film's level (propagation). Helium II can escape from any non-closed surface via creeping toward and eventually evaporating from capillaries of 10^{-7} to 10^{-8} m or greater.

Max von Laue

Max Theodor Felix von Laue, was a German physicist who won the Nobel Prize in Physics in 1914 for his discovery of the diffraction of X-rays by crystals.

In addition to his scientific endeavors with contributions in optics, crystallography, quantum theory, superconductivity, and the theory of relativity, he had a number of administrative positions which advanced and guided German scientific research and development during four decades. He was instrumental in re-establishing and organizing German science after World War II.

Born	9 October 1879, Pfaffendorf, Germany
Died	24 April 1960 (age 80), Berlin, Germany
Nationality	German
Fields	Physics
Institutions	University of Zürich, University of Frankfurt University of Berlin, Max Planck Institute
Alma mater	University of Strasbourg, University of Göttingen University of Munich, University of Berlin
Doctoral advisors	Max Planck
Doctoral students	Fritz London, Leó Szilárd, Max Kohler, Erna Weber
Known for	Diffraction of X rays
Notable awards	Nobel Prize in Physics (1914)

DISCOVERY OF X-RAYS DIFFRACTION (1914)

Laue Method

A single-crystal sample is held stationary in a beam of white X-rays. Each set of lattice planes then chooses its own wavelength to satisfy the Bragg condition. The Laue method is convenient for the rapid determination of crystal orientation and symmetry. It is also used to study the extent of crystalline imperfection under mechanical and thermal treatment. But, it is not used for crystal structure determination, because the wavelength of a particular reflection is unknown.

His best known work, however, for which he received the Nobel Prize for Physics in 1914, for his discovery of the diffraction of X-rays on crystals. This discovery originated, as he related in his Nobel Lecture, when he was discussing problems related to the passage of waves of light through a periodic, crystalline arrangement of particles. The idea then came to him that the much shorter electromagnetic rays, which X-rays were supposed to be, would cause in such a medium some kind of diffraction or interference phenomena and that a crystal would provide such a medium. Although his colleagues Sommerfeld, W. Wien and others, with whom he discussed the idea on a skiing expedition, raised objections to the idea, W. Friedrich, one of Sommerfeld's assistants and P. Knipping tested it out experimentally and, after some failures, succeeded in proving it to be correct. Von Laue worked out the mathematical formulation of it and the discovery was published in 1912. It established the fact that X-rays are electromagnetic in nature and it opened the way to the later work of Sir William and Sir Lawrence Bragg. Subsequently von Laue made other contributions to this subject.

William Lawrence Bragg

Sir William Lawrence Bragg, was an Australian born British physicist who shared the Nobel Prize in Physics in 1915 with his father Sir William Henry Bragg. He is, till date, the youngest Nobel laureate. He was the director of the Cavendish Laboratory, Cambridge, when the epochal discovery of the structure of DNA was made by James D. Watson and Francis Crick in February 1953.

Born	31 March 1890, North Adelaide, South Australia
Died	1 July 1971 (age 81), Waldringfield, Ipswich, Suffolk, England
Nationality	United Kingdom
Institutions	University of Manchester, University of Cambridge
Alma mater	University of Adelaide, University of Cambridge
Doctoral advisors	J. J. Thomson, W.H. Bragg
Doctoral students	John Crank, Ronald Wilfried Gurney
Known for	X-ray diffraction, Bragg's law
Notable awards	Nobel Prize in Physics (1915)
At 25, the youngest person ever to receive a Nobel Prize. He was the son of W.H. Bragg. J. J. Thomson and W.H. Bragg were his equivalent mentors	

BRAGG'S LAW (1915)

In physics, Bragg's law gives the angles for coherent and incoherent scattering from a crystal lattice. When X-rays are incident on an atom, they make the electronic cloud move as does any electromagnetic wave. The movement of these charges re-radiates waves with the same frequency (blurred slightly due to a variety of effects); this phenomenon is known as Rayleigh scattering (or elastic scattering). The scattered waves can themselves be scattered but this secondary scattering is assumed to be negligible. A similar process occurs upon scattering neutron waves from the nuclei or by a coherent spin interaction with an unpaired electron. These re-emitted wave fields interfere with each other either constructively or destructively (overlapping waves either add together to produce stronger peaks or subtract from each other to some degree), producing a diffraction pattern on a detector or film. The resulting wave interference pattern is the basis of diffraction analysis. Both neutron and X-ray wavelengths are comparable with inter-atomic distances (~150 pm) and thus are an excellent probe for this length scale.

The interference is constructive when the phase shift is a multiple of 2π; this condition can be expressed by Bragg's law,

$n\lambda = 2d \sin \theta$

where n is an integer, λ is the wavelength of incident wave, d is the spacing between the planes in the atomic lattice, and θ is the angle between the incident ray and the scattering planes. The moving particles, including electrons, protons and neutrons, have an associated de Broglie wavelength.

Bragg's law is the result of experiments into the diffraction of X-rays or neutrons off crystal surfaces at certain angles, derived by physicist Sir William Lawrence Bragg in 1912 and first presented on 11 November 1912 to the Cambridge Philosophical Society. Although simple, Bragg's law confirmed the existence of real particles at the atomic scale, as well as providing a powerful new tool for studying crystals in the form of X-ray and neutron diffraction. William Lawrence Bragg and his father, Sir William Henry Bragg, were awarded the Nobel Prize in physics in 1915 for their work in determining crystal structures beginning with NaCl, ZnS, and diamond.

William Henry Bragg

Sir William Henry Bragg, was a British physicist, chemist, mathematician and active sportsman who uniquely shared a Nobel Prize with his son William Lawrence Bragg–the 1915 Nobel Prize in Physics. The mineral Braggite is named after him and his son.

Born	2 July 1862, Wigton, Cumberland, England
Died	10 March 1942 (age 79), London, England
Nationality	British
Residence	England
Fields	Physicist
Institutions	University of Adelaide, University of Leeds University College London, Royal Institution
Alma mater	Cambridge University
Academic advisors	J. J. Thomson
Doctoral students	W. L. Bragg, Kathleen Lonsdale, William Thomas Astbury
Notable students	John Burton Cleland
Known for	X-ray diffraction
Notable awards	Nobel Prize in Physics (1915)
He is the father of William Lawrence Bragg. Father and son jointly won the Nobel Prize	

BRAGG'S LAW (1915)

Bragg diffraction (also referred to as the Bragg formulation of X-ray diffraction) was first proposed by William Lawrence Bragg and William Henry Bragg in 1913 in response to their discovery that crystalline solids produced surprising patterns of reflected X-rays (in contrast to that of, say, a liquid). They found that in these crystals, for certain specific wavelengths and incident angles, intense peaks of reflected radiation (known as *Bragg peaks*) were produced. The concept of *Bragg diffraction* applies equally to neutron diffraction and electron diffraction processes.

W. L. Bragg explained this result by modeling the crystal as a set of discrete parallel planes separated by a constant parameter d. It was proposed that the incident X-ray radiation would produce a Bragg peak if their reflections off the various planes interfered constructively.

Bragg diffraction occurs when electromagnetic radiation or subatomic particle waves with wavelength comparable to atomic spacings are incident upon a crystalline sample, scattered in a specular fashion by the atoms in the system, and undergo constructive interference in accordance to Bragg's law. For a crystalline solid, the waves are scattered from lattice planes separated by the interplanar distance d. Where the scattered waves interfere constructively; they remain in phase since the path length of each wave is equal to an integer multiple of the wavelength. The path difference between two waves undergoing constructive interference is given by $2d \sin \theta$, where θ is the scattering angle. This leads to Bragg's law which describes the condition for constructive interference from successive crystallographic planes (h, k, l) of the crystalline lattice: $2d \sin \theta = n\lambda$

where n is an integer determined by the order given, and λ is the wavelength. A diffraction pattern is obtained by measuring the intensity of scattered waves as a function of scattering angle. Very strong intensities known as Bragg peaks are obtained in the diffraction pattern when scattered waves satisfy the Bragg condition.

Charles Glover Barkla

Charles Glover Barkla, was British physicist, and the winner of the Nobel Prize in Physics in 1917 for his work in X-ray spectroscopy and related areas in the study of X-rays (Roentgen rays).

Born	27 June 1877, Widnes, Cheshire, England
Died	23 October 1944 (age 67), Edinburgh, Scotland
Nationality	United Kingdom
Fields	Physics
Institutions	Cambridge University, Liverpool University, King's College London, University of Edinburgh
Alma mater	University College Liverpool, Cambridge University
Doctoral advisors	J. J. Thomson, Oliver Lodge
Known for	X-ray scattering, X-ray spectroscopy
Notable awards	Nobel Prize in Physics (1917) Hughes Medal of the Royal Society

X-RAY EMISSION SPECTROSCOPY (1917)

X-ray scattering techniques are a family of non-destructive analytical techniques which reveal information about the crystallographic structure, chemical composition, and physical properties of materials and thin films. These techniques are based on observing the scattered intensity of an X-ray beam hitting a sample as a function of incident and scattered angle, polarization, and wavelength or energy.

Karl Manne Georg Siegbahn from Uppsala, Sweden (Nobel Prize 1924), painstakingly produced numerous diamond-ruled glass diffraction gratings for his spectrometers, was one of the pioneers in developing X-ray emission spectroscopy (also called X-ray fluorescence spectroscopy). He measured the X-ray wavelengths of many elements to high precision, using high-energy electrons as excitation source.

Intense and wavelength-tunable X-rays are now typically generated with synchrotrons. In a material, the X-rays may suffer an energy loss compared to the incoming beam. This energy loss of the re-emerging beam reflects an internal excitation of the atomic system, an X-ray analogue to the well-known Raman spectroscopy that is widely used in the optical region.

In the X-ray region there is sufficient energy to probe changes in the electronic state (transitions between orbitals; this is in contrast with the optical region, where the energy loss is often due to changes in the state of the rotational or vibrational degrees of freedom). For instance, in the ultra soft X-ray region (below about 1 keV), crystal field excitations give rise to the energy loss.

The photon-in-photon-out process may be thought of as a scattering event. When the X-ray energy corresponds to the binding energy of a core-level electron, this scattering process is resonantly enhanced by many orders of magnitude. This type of X-ray emission spectroscopy is often referred to as resonant inelastic X-ray scattering (RIXS).

Due to the wide separation of orbital energies of the core levels, it is possible to select a certain atom of interest. The small spatial extent of core level orbitals forces the RIXS process to reflect the electronic structure in close vicinity of the chosen atom. Thus RIXS experiments give valuable information about the local electronic structure of complex systems, and theoretical calculations are relatively simple to perform.

Max Planck

Max Planck, was a German physicist. He is considered to be the founder of the quantum theory, and thus one of the most important physicists of the twentieth century. Planck was awarded the Nobel Prize in Physics in 1918.

Born	23 April 1858, Kiel, Holstein
Died	4 October 1947 (age 89), Göttingen, West Germany
Nationality	German
Fields	Physics
Institutions	University of Kiel, University of Berlin, University of Göttingen, Kaiser-Wilhelm-Gesellschaft
Alma mater	Ludwig Maximilian, University of Munich
Doctoral advisors	Alexander von Brill
Doctoral students	Gustav Ludwig Hertz, Erich Kretschmann, Walther Meifsner, Walter Schottky, Max von Laue, Max Abraham, Moritz Schlick, Walther Bothe, Julius Edgar Lilienfeld
Known for	Planck constant, Planck postulate, Planck's law of black-body radiation
Notable awards	Nobel Prize in Physics (1918)
He is the father of Erwin Planck who was executed in 1945 by the Gestapo for his part in the July 20 plot	

PLANCK'S LAW (1918)

In 1894 Planck turned his attention to the problem of black-body radiation. He had been commissioned by electric companies to create maximum light from light bulbs with minimum energy. The problem had been stated by Kirchhoff in 1859: how does the intensity of the electromagnetic radiation emitted by a black-body depend on the frequency of the radiation and the temperature of the body? The question had been explored experimentally, but no theoretical treatment was agreed with experimental values. Wilhelm Wien proposed Wien's law, which correctly predicted the behavior at high frequencies, but failed at low frequencies. The Rayleigh-Jeans law, another approach to the problem, created what was later known as the ultraviolet catastrophe, but contrary to many textbooks this was not a motivation for Planck. Planck's first proposed solution to the problem in 1899 followed from what Planck called the principle of elementary disorder, which allowed him to derive Wien's law from a number of assumptions about the entropy of an ideal oscillator, creating what was referred-to as the Wien-Planck law. Soon it was found that experimental evidence did not confirm the new law at all, to Planck's frustration. Planck revised his approach, deriving the first version of the famous Planck black-body radiation law, which described the experimentally observed black-body spectrum well. It was first proposed in a meeting of the DPG on 19 October, 1900 and published in 1901. This first derivation did not include energy quantization, and did not use statistical mechanics, to which he held an aversion. In November 1900, Planck revised this first approach, relying on Boltzmann's statistical interpretation of the second law of thermodynamics, as a way of gaining a more fundamental understanding of the principles behind his radiation law. As Planck was deeply suspicious of the philosophical and physical implications of such an interpretation of Boltzmann's approach, his recourse to them was, as he later put it, an act of despair ... I was ready to sacrifice any of my previous convictions about physics. The central assumption behind his new derivation, presented to the DPG on 14 December 1900, was the supposition, now known as the Planck postulate, that electromagnetic energy could be emitted only in quantized form, in other words, the energy could only be a multiple of an elementary unit $E = h\gamma$, where h is Planck's constant, also known as Planck's action quantum and γ is the frequency of the radiation.

Johannes Stark

Johannes Stark, was a German physicist, and Physics Nobel Prize laureate who was closely involved with the Deutsche Physik movement under the Nazi regime.

Born	15 April 1874, Schickenhof, German Empire
Died	21 June 1957 (age 83), Traunstein, West Germany
Nationality	Germany
Fields	Physics
Institutions	University of Göttingen, Technische Hochschule, Hannover Technische Hochschule, Aachen, University of Greifswald, University of Würzburg
Alma mater	University of Munich
Doctoral advisors	Eugen von Lommel
Known for	Stark effect
Notable awards	Nobel Prize in Physics (1919)

Stark's scientific works cover three large fields: the electric currents in gases, spectroscopic analysis, and chemical valency. His spectroscopic work deals with the connection between the alteration in the structure and in the spectrum of chemical atoms. In 1919 Stark was awarded the Nobel Prize for Physics for his discovery of the Doppler effect in canal rays and the splitting of spectral lines in electric fields. The prize enabled him to set up his own private laboratory.

STARK EFFECT (1919)

The Stark effect is the shifting and splitting of spectral lines of atoms and molecules due to the presence of an external static electric field. The amount of splitting and or shifting is called the Stark splitting or Stark shift. In general one distinguishes first and second order Stark effects. The first-order effect is linear in the applied electric field, while the second-order effect is quadratic in the field.

The Stark effect is responsible for the pressure broadening of spectral lines by charged particles. When the split/shifted lines appear in absorption, the effect is called the inverse Stark effect.

The Stark effect is the electric analogue of the Zeeman effect where a spectral line is split into several components due to the presence of a magnetic field.

The Stark effect can be explained with fully quantum mechanical approaches, but it has also been a fertile testing ground for semi-classical methods.

The effect is named after Johannes Stark, who discovered it in 1913. It was independently discovered in the same year by the Italian physicist Antonino Lo Surdo, and in Italy it is thus sometimes called the Stark-Lo Surdo effect. The discovery of this effect contributed importantly to the development of quantum theory. Ironically, soon after their discoveries, both Stark and Lo Surdo rejected developments in modern physics and allied themselves with the political and racial programs of Hitler and Mussolini.

Inspired by the magnetic Zeeman effect, and especially by Lorentz' explanation of it, Woldemar Voigt performed classical mechanical calculations of quasi-elastically bound electrons in an electric field. By using experimental indices of refraction he gave an estimate of the Stark splittings. This estimate was a few orders of magnitude too low. Not deterred by this prediction, Stark undertook measurements on excited states of the hydrogen atom and succeeded in observing splittings.

Charles Édouard Guillaume

Charles Édouard Guillaume, was a Swiss physicist who received the Nobel Prize in Physics in 1920 in recognition of the service he had rendered to precision measurements in physics by his discovery of anomalies in nickel steel alloys.

Born	15 February 1861, Fleurier, Switzerland
Died	13 May 1938 (age 77), Sèvres, France
Nationality	Swiss
Fields	Physics
Institutions	Bureau International des Poids et Mesures, Sèvres
Alma mater	ETH Zurich
Known for	Invar and Elinvar
Notable awards	Nobel Prize in Physics (1920)

INVAR AND ELINVAR (1920)

Invar: Invar, also known generically as FeNi36, is a nickel steel alloy notable for its uniquely low coefficient of thermal expansion. It was invented in 1896 by Swiss scientist Charles Édouard Guillaume. He received the Nobel Prize in Physics in 1920 for this discovery, which shows the importance of this alloy in scientific instruments. Invar is a registered trademark of ArcelorMittal–Stainless and Nickel Alloys, formerly known as Imphy Alloys, however FeNi36 is also manufactured by Japanese companies. Like other nickel/iron compositions, Invar is a solid solution; that is, it is a single-phase alloy, similar to a dilution of common table salt mixed into water. The name Invar comes from the word invariable, referring to its lack of expansion or contraction with temperature changes.

Common grades of Invar have an α (20–100°C) of about 1.2×10^{-6} K^{-1} (1.2 ppm/°C). However, extra-pure grades (<0.1% Co) can readily produce values as low as 0.62–0.65 ppm/°C. Some formulations display negative thermal expansion (NTE) characteristics. It is used where high dimensional stability is required, such as precision instruments, clocks, seismic creep gauges, television shadow-mask frames, valves in motors, and antimagnetic watches. However, it has a propensity to creep. In land surveying, when first-order (high-precision) elevation leveling is to be performed, the leveling rods used are made of Invar, instead of wood, fiberglass, or other metals.

Elinvar: Elinvar is a nickel steel alloy with a modulus of elasticity which does not change much with temperature changes. The name is a contraction of the French Elasticité invariable. It was invented around the 1920s by Charles Édouard Guillaume, a Swiss physicist who also invented Invar, another alloy of nickel and iron, which has very low thermal expansion. Guillaume won the 1920 Nobel Prize in Physics for these discoveries, which indicates how important these alloys were for scientific instruments. Elinvar consists of 59% iron, 36% nickel, and 5% chromium. It is almost nonmagnetic and corrosion resistant.

The largest use of Elinvar was in balance springs for mechanical watches and chronometers. A major cause of inaccuracy in watches and clocks was that ordinary steels used in springs lost elasticity slightly as the temperature increased, so the balance wheel would oscillate more slowly back and forth, and the clock would lose time. Chronometers and precision watches required complex temperature–compensated balance wheels for accurate timekeeping. Springs made of Elinvar, and other low temperature coefficient alloys such as Nivarox that followed, were not affected by temperature, so they made the temperature-compensated balance wheel obsolete.

Albert Einstein

 Albert Einstein, was a theoretical physicist, philosopher and author who is widely regarded as one of the most influential and best known scientists and intellectuals of all time. He is often regarded as the father of modern physics. He received the 1921 Nobel Prize in Physics for his services to Theoretical Physics, and especially for his discovery of the law of the photoelectric effect.

Born	14 March 1879, Ulm, Kingdom of Württemberg, German Empire
Died	18 April 1955 (age 76), Princeton, New Jersey, USA Resting place Grounds of the Institute for Advanced Study, Princeton, New Jersey
Residence	Germany, Italy, Switzerland, USA,
Ethnicity	Jewish
Citizenship	Württemberg/Germany (until 1896), Stateless (1896–1901), Switzerland (from 1901), Austria (1911–12), Germany (1914–33), United States (from 1940)
Alma mater	ETH Zurich, University of Zurich
Known for	General relativity, Special relativity, Photoelectric effect, Brownian motion, Mass-energy equivalence, Einstein field equations, Unified Field Theory, Bose–Einstein statistics
Notable awards	Nobel Prize in Physics (1921), Copley Medal (1925) Max Planck Medal (1929), Time Person of the Century

PHOTOELECTRIC EFFECT (1921)

The photoelectric effect is a phenomenon in which electrons are emitted from matter (metals and non-metallic solids, liquids or gases) as a consequence of their absorption of energy from electromagnetic radiation of very short wavelength, such as visible or ultraviolet light. Electrons emitted in this manner may be referred to as photoelectrons. As it was first observed by Heinrich Hertz in 1887, the phenomenon is also known as the Hertz effect, although the latter term has fallen out of general use. Hertz observed and then showed that electrodes illuminated with ultraviolet light create electric sparks more easily.

The photoelectric effect takes place with photon energies from about a few electron volts to, in high atomic number elements, over 1 Mev. At the high photon energies comparable to the electron rest energy of 511 keV, Compton scattering, another process, may take place, and above twice this (1.022 Mev) pair production may take place.

Study of the photoelectric effect led to important steps in understanding the quantum nature of light and electrons and influence the formation of the concept of wave-particle duality.

The term may also, but incorrectly, refer to related phenomena such as the photoconductive effect (also known as photoconductivity or photo resistivity), the photovoltaic effect, or the photo electrochemical effect which are, in fact, distinctly different.

Other important works are, General relativity, Special relativity, Brownian motion, Mass-energy equivalence, Einstein field equations, Unified Field Theory, Bose-Einstein statistics.

Niels Bohr

Niels Henrik David Bohr, was a Danish physicist who made foundational contributions of understanding atomic structure and quantum mechanics, for which he received the Nobel Prize in Physics in 1922.

Bohr mentored and was collaborated with many of the top physicists of the century at his institute in Copenhagen. He was part of a team of physicists working on the Manhattan Project. Bohr married Margrethe Nørlund in 1912, and one of their son, Aage Bohr, grew up to be an important physicist, who in 1975 also received the Nobel Prize. Bohr has been described as one of the most influential physicists of the 20th century.

Born	7 October 1885, Copenhagen, Denmark
Died	18 November 1962 (age 77), Copenhagen, Denmark
Nationality	Denmark
Fields	Physics
Institutions	University of Copenhagen
Alma mater	University of Copenhagen
Doctoral advisor	Christian Christiansen
Academic advisors	J. J. Thomson, Ernest Rutherford
Doctoral students	Hendrik Anthony Kramers
Known for	Copenhagen interpretation, Complementarity, Bohr model, Sommerfeld-Bohr theory, BKS theory, Bohr-Einstein debates
Influences	Ernest Rutherford
Influenced	Werner Heisenberg, Wolfgang Pauli, Paul Dirac, Lise Meitner, Max Delbrück
Notable awards	Nobel Prize in Physics (1922)

BOHR MODEL (1922)

In atomic physics, the Bohr model, devised by Niels Bohr, depicts the atom as a small, positively charged nucleus surrounded by electrons that travel in circular orbits around the nucleus, similar in structure to the solar system, but with electrostatic forces providing attraction, rather than gravity. This was an improvement on the earlier cubic model (1902), the plum-pudding model (1904), the Saturnian model (1904), and the Rutherford model (1911). Since the Bohr model is a quantum physics-based modification of the Rutherford model, many sources combine the two, referring to the Rutherford-Bohr model.

Introduced by Niels Bohr in 1913, the model's key success lay in explaining the Rydberg formula for the spectral emission lines of atomic hydrogen. While the Rydberg formula had been known experimentally, it did not gain a theoretical underpinning until the Bohr model was introduced. Not only did the Bohr model explain the reason for the structure of the Rydberg formula, it also provided a justification for its empirical results in terms of fundamental physical constants.

The Bohr model is a primitive model of the hydrogen atom. As a theory, it can be derived as a first-order approximation of the hydrogen atom using the broader and much more accurate quantum mechanics, and thus may be considered to be an obsolete scientific theory. However, because of its simplicity, and its correct results for selected systems, the Bohr model is still commonly taught to introduce students to quantum mechanics, before moving on to the more accurate but more complex valence shell atom. A related model was originally proposed by Arthur Erich Haas in 1910, but was rejected. The quantum theory of the period between Planck's discovery of the quantum (1900) and the advent of a full-blown quantum mechanics (1925) is often referred to as the old quantum theory.

Robert Andrews Millikan

 Robert A. Millikan, was an American experimental physicist, and Nobel laureate in physics for his measurement of the charge on the electron and for his work on the photoelectric effect. He served as a president of Caltech from 1921 to 1945.

Born	22 March 1868, Morrison, Illinois, USA
Died	19 December 1953 (age 85), San Marino, California, USA
Nationality	United States
Fields	Physics
Institutions	University of Chicago, California Institute of Technology
Alma mater	Oberlin College, Columbia University
Doctoral advisors	Michael I. Pupin, Albert Michelson
Doctoral students	William Pickering, Robley D. Evans, Harvey Fletcher Chung-Yao Chao
Known for	Charge of the electron, Advanced cosmic ray physics
Notable awards	Nobel Prize in Physics (1923), Franklin Medal (1937)

ELEMENTARY CHARGE (1923)

The elementary charge, usually denoted as e or sometimes q, is the electric charge carried by a single proton, or equivalently, the absolute value of the electric charge carried by a single electron. This elementary charge is a fundamental physical constant. To avoid confusion over its sign, e is sometimes called the elementary positive charge. This charge has a measured value of approximately $1.60217648 7 \times 10^{19}$ coulombs. In the CGS system, the value for e is $4.80320427(12) \times 10^{-10}$ stat coulombs. In the system of atomic units as well as some other systems of natural units, e functions as the unit of electric charge, i.e. $e = 1$ in those unit systems.

The magnitude of the elementary charge was first measured in Robert A. Millikan's noted oil-drop experiment in 1909.

Charge quantization is the statement that every stable and independent object (meaning an object that can exist independently for a prolonged period of time) has a charge which is an integer multiple of the elementary charge e: A charge can be exactly 0, or exactly e, $-e$, $2e$, etc., but not, say, $\frac{1}{2}e$, or $-3.8e$, etc. This statement must not be interpreted to include quarks or quasi particles, since neither quarks nor quasi particles possess the ability to exist on their own for prolonged periods of time. Quarks have charges that are integer multiples of $\frac{1}{3}e$.

This is the reason for the terminology elementary charge: it is meant to imply that it is an indivisible unit of charge.

The German physicist Johann Wilhelm Hittorf undertook the study of electrical conductivity in rarefied gases. In 1869, he discovered a glow emitted from the cathode that increased in size with decrease in gas pressure. In 1876, the German physicist Eugen Goldstein showed that the rays from this glow cast a shadow, and he dubbed them cathode rays. During the 1870s, the English chemist and physicist Sir William Crookes developed the first cathode ray tube to have a high vacuum inside. He then showed that the luminescence rays appearing within the tube carried energy and moved from the cathode to the anode. Furthermore, by applying a magnetic field, he was able to deflect the rays, thereby demonstrating that the beam behaved as though it were negatively charged. In 1879, he proposed that these properties could be explained by what he termed radiant matter. He suggested that this was a fourth state of matter, consisting of negatively charged molecules that were being projected with high velocity from the cathode.

Manne Siegbahn

Karl Manne Georg Siegbahn, was a Swedish physicist, and Nobel laureate in physics, for his discoveries and research in the field of X-ray spectroscopy.

He was born in Örebro, Sweden. Siegbahn obtained his Ph.D. at the Lund University in 1911; his thesis was titled Magnetische Feldmessungen (magnetic field measurements). He was acting professor for Janne Rydberg when his health was failing, and succeeded him as full professor in 1915.

Born	3 December 1886, Örebro, Sweden
Died	26 September 1978 (age 91), Stockholm, Sweden
Nationality	Swedish
Fields	Physics
Institutions	University of Lund, University of Uppsala, University of Stockholm
Alma mater	University of Lund
Known for	X-ray spectroscopy
Notable awards	Nobel Prize in Physics (1924)
He is the father of Nobel laureate Kai M. Siegbahn	

X-RAY EMISSION SPECTROSCOPY (1924)

Karl Manne Georg Siegbahn from Uppsala, Sweden, painstakingly produced numerous diamond-ruled glass diffraction gratings for his spectrometers, was one of the pioneers in developing X-ray emission spectroscopy also called X-ray fluorescence spectroscopy. He measured the X-ray wavelengths of many elements to high precision, using high-energy electrons as excitation source. Intense and wavelength-tunable X-rays are now typically generated with synchrotrons. In a material, the X-rays may suffer an energy loss compared to the incoming beam. This energy loss of the re-emerging beam reflects an internal excitation of the atomic system, an X-ray analogue to the well-known Raman spectroscopy that is widely used in the optical region.

In the X-ray region there is sufficient energy to probe changes in the electronic state. For instance, in the ultra soft X-ray region (below about 1 keV), crystal field excitations give rise to the energy loss. The photon-in-photon-out process may be thought of as a scattering event. When the X-ray energy corresponds to the binding energy of a core-level electron, this scattering process is resonantly enhanced by many orders of magnitude. This type of X-ray emission spectroscopy is often referred to as resonant inelastic X-ray scattering (RIXS). Due to the wide separation of orbital energies of the core levels, it is possible to select a certain atom of interest. The small spatial extent of core level orbital forces the RIXS process to reflect the electronic structure in close vicinity of the chosen atom. Thus RIXS experiments give valuable information about the local electronic structure of complex systems, and theoretical calculations are relatively simple to perform.

A typical XPS spectrum is a plot of the number of electrons detected (sometimes per unit time) versus the binding energy of the electrons detected (X-axis, abscissa). Each element produces a characteristic set of XPS peaks at characteristic binding energy values that directly identify each element that exist in or on the surface of the material being analyzed. These characteristic peaks correspond to the electron configuration of the electrons within the atoms, e.g. 1s, 2s, 2p, 3s, etc. The number of detected electrons in each of the characteristic peaks is directly related to the amount of element within the area (volume) irradiated. To generate atomic percentage values, each raw XPS signal must be corrected by dividing its signal intensity (number of electrons detected) by a relative sensitivity factor (RSF) and normalized over all of the elements detected.

James Franck

 James Franck, was a German Jewish physicist and Nobel laureate. In 1925, Franck received the Nobel Prize in Physics, for his work in 1912–1914, which included the Franck-Hertz experiment, an important confirmation of the Bohr model of the atom.

Born	26 August 1882, Hamburg, German Empire
Died	21 May 1964 (age 81), Göttingen, West Germany
Nationality	German Jew
Fields	Physics
Institutions	University of Berlin, University of Göttingen, Johns Hopkins University, University of Chicago
Alma mater	University of Heidelberg, University of Berlin
Doctoral advisor	Emil Gabriel Warburg
Doctoral students	Wilhelm Hanle, Arthur R. von Hippel
Known for	Franck-Condon principle, Franck-Hertz experiment
Notable awards	Nobel Prize in Physics (1925)

FRANCK-HERTZ EXPERIMENT (1925)

The Franck-Hertz experiment was a physics experiment that provided support for the Bohr model of the atom, a precursor to quantum mechanics. In 1914, the German physicists James Franck and Gustav Ludwig Hertz sought to experimentally probe the energy levels of the atom. The famous Franck-Hertz experiment elegantly supported Niels Bohr's model of the atom, with electrons orbiting the nucleus with specific, discrete energies. Franck and Hertz were awarded the Nobel Prize in Physics in 1925 for this work.

The Franck-Hertz experiment confirmed Bohr's quantized model of the atom by demonstrating that atoms could indeed only absorb (and be excited by) specific amounts of energy (quanta).

Franck and Hertz were able to explain their experiment in terms of elastic and inelastic collisions. At low potentials, the accelerated electrons acquired only a modest amount of kinetic energy. When they encountered mercury atoms in the tube, they participated in purely elastic collisions. This is due to the prediction of quantum mechanics that an atom can absorb no energy until the collision energy exceeds that required to lift an electron into a higher energy state.

With purely elastic collisions, the total amount of kinetic energy in the system remains the same. Since electrons are over one thousand times less massive than even the lightest atoms, this meant that the electrons held on to the vast majority of that kinetic energy. Higher potentials served to drive more electrons through the grid to the anode and increase the observed current, until the accelerating potential reached 4.9 volts.

The classic experiment involved a tube containing low pressure gas fitted with three electrodes: an electron emitting cathode, a mesh grid for acceleration, and an anode. The anode was held at a slightly negative electrical potential relative to the grid, so that electrons had to have a small amount of kinetic energy to reach it after passing the grid. Instruments were fitted to measure the current passing between the electrodes, and to adjust the potential difference between the cathode and the accelerating grid.

- At low potential differences up to 4.9 volts when the tube contained mercury vapor, the current through the tube increased steadily with increasing potential difference. The higher voltage increase the electric field in the tube and electrons were drawn more forcefully towards and through the accelerating grid.
- At 4.9 volts the current drops sharply, almost back to zero.
- The current increase steadily once again if the voltage is increased further, until 9.8 volts is reached (exactly 4.9 + 4.9 volts).
- At 9.8 volts a similar sharp drop is observed.
- This series of dips in current at approximately 4.9 volt increments will visibly continue to potentials of at least 100 volts.

Gustav Ludwig Hertz

 Gustav Ludwig Hertz, was a German Jew experimental physicist and Nobel Prize winner. He received his doctorate in 1911 under Heinrich Leopold Rubens.

From 1911 to 1914, Hertz was an assistant to Rubens at the University of Berlin. It was during this time that Hertz and James Franck performed experiments on inelastic electron collisions in gases, known as the Franck-Hertz experiments, and for which they received the Nobel Prize in Physics in 1925.

Born	22 July 1887, Hamburg, Germany
Died	30 October 1975 (age 88), Berlin, Germany
Nationality	German Jew
Fields	Physics
Institutions	Halle University
Alma mater	Humboldt University of Berlin
Doctoral advisor	Heinrich Rubens, Max Planck
Known for	Franck-Hertz experiment
Notable awards	Nobel Prize in Physics (1925)
Father of Carl Hellmuth Hertz	

FRANCK-HERTZ EXPERIMENT (1925)

In 1914, James Franck and Gustav Hertz performed an experiment which demonstrated the existence of excited states in mercury atoms, helping to confirm the quantum theory which predicted that electrons occupied only discrete, quantized energy states. Electrons were accelerated by a voltage toward a positively charged grid in a glass envelope filled with mercury vapor. Past the grid was a collection plate held at a small negative voltage with respect to the grid. The values of accelerating voltage where the current dropped gave a measure of the energy necessary to force an electron to an excited state.

The lowest energy electronic excitation a mercury atom can participate in requires 4.9 electron volts (eV). When the accelerating potential reached 4.9 volts, each free electron possessed exactly 4.9 eV of kinetic energy (above its rest energy at that temperature) when it reached the grid. Consequently, a collision between a mercury atom and a free electron at that point could be inelastic: that is, a free electron's kinetic energy could be converted into potential energy by raising the energy level of an electron bound to a mercury atom, this is called exciting the mercury atom. With the loss of all its acquired kinetic energy in this way, the free electron can no longer overcome the slight negative potential at the ground electrode, and the measured current drops sharply.

As the voltage is increased, electrons will participate in one inelastic collision, lose their 4.9 eV, but then continue to be accelerated. In this manner, the current rises again after the accelerating potential exceeds 4.9 V. At 9.8 V, the situation changes again. There, each electron now has just enough energy to participate in *two* inelastic collisions, excite two mercury atoms, and then be left with no kinetic energy. Once again, the observed current drops. At intervals of 4.9 volts this process will repeat; each time the electrons will undergo one additional inelastic collision.

New findings of this experiment have found that the spacing between minima and maxima increase with number of minima and vary with temperature.

Jean Baptiste Perrin

Jean Baptiste Perrin, was a French physicist and Nobel laureate. Jean Perrin received the Nobel Prize in Physics in 1926 for this and other work on the discontinuous structure of matter, which put a definite end to the long struggle regarding the question of the physical reality of molecules.

Born	30 September 1870, Lille, France
Died	17 April 1942 (age 71), New York City, USA
Nationality	France
Fields	Physics
Institutions	École Normale Supérieure, University of Paris
Alma mater	École Normale Supérieure
Known for	Nature of cathode rays, Brownian motion
Notable awards	Nobel Prize in Physics (1926)

BROWNIAN MOTION (1926)

Brownian motion (named after the Scottish botanist Robert Brown) or pedesis is the seemingly random movement of particles suspended in a fluid (i.e. a liquid such as water or air) or the mathematical model used to describe such random movements, often called a particle theory.

The mathematical model of Brownian motion has several real-world applications. An often quoted example is stock market fluctuations. However, movements in share prices may arise due to unforeseen events which do not repeat themselves, and physical and economic phenomena are not comparable.

Brownian motion is among the simplest of the continuous-time stochastic (or random) processes, and it is a limit of both simpler and more complicated stochastic processes. This universality is closely related to the universality of the normal distribution. In both cases, it is often mathematical convenience rather than the accuracy of the models that motivates their use.

The Roman Lucretius's scientific poem 'On the Nature of Things' has a remarkable description of Brownian motion of dust particles. He uses this as a proof of the existence of atoms.

Observe what happens when Sunbeams are admitted into a building and shed light on its shadowy places. You will see a multitude of tiny particles mingling in a multitude of ways. Their dancing is an actual indication of underlying movements of matter that are hidden from our sight. It originates with the atoms which move of themselves [i.e. spontaneously]. Then those small compound bodies that are least removed from the impetus of the atoms are set in motion by the impact of their invisible blows and in turn cannon against slightly larger bodies. So the movement mounts up from the atoms and gradually emerges to the level of our senses, so that those bodies are in motion that we see in sunbeams, moved by blows that remain invisible.

Although the mingling motion of dust particles is caused largely by air currents, the glittering, tumbling motion of small dust particles is, indeed, caused chiefly by true Brownian dynamics.

The first person to describe the mathematics behind Brownian motion was Thorvald N. Thiele in 1880 in a paper on the method of least squares. This was followed independently by Louis Bachelier in 1900 in his PhD thesis The theory of speculation, in which he presented a stochastic analysis of the stock and option markets. However, it was Albert Einstein and Marian Smoluchowski (1906) who independently brought the solution of the problem to the attention of physicists, and presented it as a way to indirectly confirm the existence of atoms and molecules.

Arthur Compton

 Arthur Holly Compton, was an American physicist and Nobel laureate in physics for his discovery of the Compton Effect. He served as Chancellor of Washington University in St. Louis from 1945 to 1953.

Compton was awarded the Nobel Prize in Physics in 1927. He developed the method for observing at the same instant individual scattered X-ray photons and the recoil electrons. In Germany, Walther Bothe and Hans Geiger independently developed a similar method.

Born	10 September 1892, Wooster, Ohio, USA
Died	15 March 1962 (age 69), Berkeley, California, USA
Nationality	United States
Fields	Physics
Institutions	Washington University in St. Louis, University of Chicago, University of Minnesota
Alma mater	College of Wooster Princeton University
Doctoral advisor	Owen Willans Richardson, H. L. Cooke
Doctoral students	Winston H. Bostick, Robert S. Shankland
Known for	Compton effect, Compton length, Compton scattering Compton wavelength, Compton shift
Notable awards	Nobel Prize in Physics (1927), Franklin Medal (1940)

COMPTON SCATTERING (1927)

In physics, Compton scattering is a type of scattering that X-rays and gamma rays undergo in matter. The inelastic scattering of photons in matter results in a decrease in energy of an X-ray or gamma ray photon, called the Compton effect. Part of the energy of the X/gamma ray is transferred to a scattering electron, which recoils and is ejected from its atom, and the rest of the energy is taken by the scattered, degraded photon.

Inverse Compton scattering also exists, where the photon gains energy upon interaction with matter. Since the wavelength of the scattered light is different from the incident radiation, Compton scattering is an example of inelastic scattering, but the origin of the effect can be considered as an elastic collision between a photon and an electron. The amount the wavelength changes by is called the Compton shift. Although nuclear Compton scattering exists, Compton scattering usually refers to the interaction involving only the electrons of an atom. The Compton effect was observed by Arthur Holly Compton in 1923 at Washington University in St. Louis and further verified by his graduate student Y. H. Woo in the years following. Compton earned the 1927 Nobel Prize in Physics for the discovery.

The effect is important because it demonstrates that light cannot be explained purely as a wave phenomenon. Thomson scattering, the classical theory of an electromagnetic wave scattered by charged particles, cannot explain low intensity shifts in wavelength. Classically, light of sufficient intensity for the electric field to accelerate a charged particle to a relativistic speed will cause radiation-pressure recoil and an associated Doppler shift of the scattered light, but the effect would become arbitrarily small at sufficiently low light intensities regardless of wavelength. Light must behave as if it consists of particles in order to explain the low-intensity Compton scattering. Compton's experiment convinced physicists that light can behave as a stream of particle-like objects (quanta) whose energy is proportional to the frequency.

The interaction between electrons and high energy photons comparable to the rest energy of the electron, 511 keV results in the electron being given part of the energy, and a photon containing the remaining energy being emitted in a different direction from the original, so that the overall momentum of the system is conserved. If the photon still has enough energy left, the process may be repeated.

Charles Thomson Rees Wilson

Charles Thomson Rees Wilson, was a Scottish physicist and meteorologist who received the Nobel Prize in Physics in 1927 for his invention of the Cloud Chamber.

Born	14 February 1869, Midlothian, Scotland
Died	15 November 1959 (age 90), Edinburgh, Scotland
Nationality	Scottish
Fields	Physics
Institutions	University of Cambridge
Alma mater	University of Manchester, University of Cambridge
Academic advisor	J. J. Thomson
Doctoral students	Cecil Frank Powell
Known for	Cloud chamber
Notable awards	Nobel Prize in Physics (1927)

CLOUD CHAMBER (1927)

The cloud chamber, also known as the Wilson chamber, is used for detecting particles of ionizing radiation. In its most basic form, a cloud chamber is a sealed environment containing a super cooled, supersaturated water or alcohol vapor. When an alpha particle or beta particle interacts with the mixture, it ionizes it. The resulting ions act as condensation nuclei, around which a mist will form (because the mixture is on the point of condensation). The high energies of alpha and beta particles mean that a trail is left, due to many ions being produced along the path of the charged particle. These tracks have distinctive shapes (for example, an alpha particle's track is broad and straight, while an electron's is thinner and shows more evidence of deflection by collisions). When any uniform magnetic field is applied across the cloud chamber, positively and negatively charged particles will curve in opposite directions, according to the Lorentz force law with two particles of opposite charge.

Charles Thomas Rees Wilson (1869–1959), a Scottish physicist, is credited with inventing the cloud chamber. Inspired by sightings of the Brocken spectre while working on the summit of Ben Nevis in 1894, he began to develop expansion chambers for studying cloud formation and optical phenomena in moist air. Very rapidly he discovered that ions could act as centers for water droplet formation in such chambers. He pursued the application of this discovery and perfected the first cloud chamber in 1911. In Wilson's original chamber the air inside the sealed device was saturated with water vapor, then a diaphragm is used to expand the air inside the chamber (adiabatic expansion). This cools the air and water vapor starts to condense. When an ionizing particle passes through the chamber, water vapor condenses on the resulting ions and the trail of the particle is visible in the vapor cloud. A diagram of Wilson's apparatus is given here. This kind of chamber is also called a pulsed chamber, because the conditions for operation are not continuously maintained. The cloud chamber was the first radioactivity detector.

Owen Willans Richardson

Sir Owen Willans Richardson, was a British physicist who won the Nobel Prize in Physics in 1928 for his work on Thermionic Emission, which lead to Richardson's Law.

Born	26 April 1879, Dewsbury, Yorkshire, England
Died	15 February 1959 (age 79), Alton, Hampshire, England
Nationality	United Kingdom
Fields	Physics
Institutions	University of Cambridge, Princeton University, King's College London
Alma mater	University of Cambridge
Doctoral advisor	J. J. Thomson
Doctoral students	Arthur Compton, Clinton Davisson
Known for	Richardson's law
Notable awards	Nobel Prize in Physics (1928), Hughes Medal (1920)

THERMIONIC EMISSION (1928)

Thermionic emission is the heat-induced flow of charge carriers from a surface or over a potential-energy barrier. This occurs because the thermal energy given to the carrier overcomes the forces restraining it. The charge carriers can be electrons or ions, and in older literature are sometimes referred to as thermions. After emission, a charge will initially be left behind in the emitting region that is equal in magnitude and opposite in sign to the total charge emitted. But if the emitter is connected to a battery, then this charge left behind will be neutralized by charge supplied by the battery, as the emitted charge carriers move away from the emitter, and finally the emitter will be in the same state as it was before emission. The thermionic emission of electrons is also known as thermal electron emission.

The classical example of thermionic emission is the emission of electrons from a hot metal cathode into a vacuum (archaically known as the Edison effect) in a vacuum tube. However, the term thermionic emission is now used to refer to any thermally excited charge emission process, even when the charge is emitted from one solid-state region into another. This process is crucially important in the operation of a variety of electronic devices and can be used for power generation or cooling. The magnitude of the charge flow increases dramatically with increasing temperature. However, vacuum emission from metals tends to become significant only for temperatures over 1000 K. The science dealing with this phenomenon has been known as thermionics, but this name seems to be gradually falling into disuse.

RICHARDSON'S LAW

In 1901 Richardson published the results of his experiments: the current from a heated wire seemed to depend exponentially on the temperature of the wire with a mathematical form similar to the Arrhenius equation. Later, he proposed that the emission law should have the mathematical form

$$J = A_G T^2 e^{\frac{-W}{kT}}$$

where J is the emission current density [SI unit: A/m^2], T is the thermodynamic temperature of the metal [SI unit: Kelvin (K)], W is the work function of the metal, k is the Boltzmann constant, and A_G is a parameter discussed below.

A_G must be written in the form

$$A_G = \lambda_R A_0$$

where λ_R is a material-specific correction factor that is typically of order 0.5, and A_0 is a universal constant given by

$$A_0 = \frac{4\pi m k^2 c}{h^3} = 1.20173 \times 10^6 \, Am^{-2}K^{-2}$$

where m and $-e$ are the mass and charge of an electron, and h is Planck's constant.

Louis de Broglie

Louis-Victor-Pierre-Raymond, 7th duc de Broglie, was a French physicist and a Nobel laureate. He was the sixteenth member elected to occupy the seat of the Académie Française in 1944, and served as Perpetual Secretary of the Académie des sciences, France.

de Broglie created a new field in physics, the Mécanique Ondulatoire, or wave mechanics, uniting the physics of light and matter. For this he won the Nobel Prize in Physics in 1929.

Born	15 August 1892, Dieppe, France
Died	19 March 1987 (age 94), Louveciennes, France
Nationality	French
Fields	Physics
Institutions	Sorbonne, University of Paris
Alma mater	Sorbonne
Doctoral advisor	Paul Langevin
Doctoral students	Jean-Pierre Vigier, Alexandru Proca
Known for	Wave nature of electrons, de Broglie wavelength
Notable awards	Nobel Prize in Physics (1929)

MATTER WAVE (1929)

In quantum mechanics, a matter wave or de Broglie wave is the wave of matter. The de Broglie relations show that the wavelength is inversely proportional to the momentum of a particle and that the frequency is directly proportional to the particle's kinetic energy. The wavelength of matter is also called de Broglie wavelength. The theory was advanced by Louis de Broglie in 1924 in his PhD thesis; he was awarded the Nobel Prize for Physics in 1929 for this work, which made him the first person to receive a Nobel Prize on a PhD thesis.

After strides made by Max Planck and Albert Einstein in understanding the behavior of electrons and what would be known as quantum physics, Niels Bohr began trying to explain how electrons behave. He came up with new fundamental ideas about electrons and mathematically derived the Rydberg equation, an equation that was discovered only through trial and error. This equation explains the energies of the light emitted when hydrogen gas is compressed and electrified. Unfortunately, his model only worked for the hydrogen-atom configuration, but his ideas were so revolutionary that they broke up the classical view of electrons' behavior and paved the way for fresh new ideas in what would become quantum physics and quantum mechanics.

Louis de Broglie tried to expand on Bohr's ideas, and he pushed for their application beyond hydrogen. In fact he looked for an equation which could explain the wavelength characteristics of all matter. His equation was experimentally confirmed in 1927 when physicists Lester Germer and Clinton Davisson fired electrons at a crystalline nickel target and the resulting diffraction pattern was found to match the predicted values. In de Broglie's equation an electron's wavelength is a function of Planck's constant divided by the object's momentum. When this momentum is very large, then an object's wavelength is very small. This is the case with everyday objects, such as a person; given the enormous momentum of a person compared with the very tiny Planck constant, the wavelength of a person would be so small as to be undetectable by any current measurement tools. On the other hand, many small particles have a very low momentum compared to macroscopic objects. In this case, the de Broglie wavelength may be large enough that the particle's wave-like nature gives observable effects.

Sir C. V. Raman

 Sir Chandrasekhara Venkata Raman, was an Indian physicist and Nobel laureate in physics recognised for his work on the molecular scattering of light and for the discovery of Raman effect, which is named after him.

On 28 February 1928, through his experiments on the scattering of light, he discovered the Raman Effect. It was instantly clear that this discovery was an important one. It gave further proof of the quantum nature of light. Raman spectroscopy came to be based on this phenomenon, and Ernest Rutherford referred to it in his presidential address to the Royal Society in 1929. Raman was president of the 16th session of the Indian Science Congress in 1929. He was conferred a knighthood, and medals and honorary doctorates by various universities. Raman was confident of winning the Nobel Prize in Physics as well, and was disappointed when the Nobel Prize went to Richardson in 1928 and to de Broglie in 1929. He did eventually win the 1930 Nobel Prize in Physics for his work on the scattering of light and for the discovery of the effect named after him.

Born	7 November 1888, Thiruvanaikoil, Tiruchirappalli, Tamil Nadu, British India
Died	21 November 1970 (aged 82), Bangalore, Karnataka, India
Nationality	Indian
Fields	Physics
Institutions	Indian Finance Department, Indian Association for the Cultivation of Science, Indian Institute of Science
Alma mater	Presidency College
Doctoral students	G. N. Ramachandran
Known for	Raman effect
Notable awards	Knight Bachelor (1929), Nobel Prize in Physics (1930) Bharat Ratna (1954), Lenin Peace Prize (1957)

He was the first Asian and first non-White to get any Nobel Prize in the Sciences. Raman also worked on the acoustics of musical instruments. He worked out the theory of transverse vibration of bowed strings, on the basis of superposition velocities. He was also the first to investigate the harmonic nature of the sound of the Indian drums such as the *tabla* and the *mridangam*.

RAMAN SCATTERING (1930)

Raman scattering or the Raman effect is the inelastic scattering of a photon, discovered by Sir Chandrasekhara Venkata Raman in liquids and by Grigory Landsberg and Leonid Mandelstam in crystals.

When light is scattered from an atom or molecule, most photons are elastically scattered (Rayleigh scattering), such that the scattered photons have the same energy (frequency) and wavelength as the incident photons. However, a small fraction of the scattered light (approximately 1 in 10 million photons) is scattered by an excitation, with the scattered photons having a frequency different from, and usually lower than, the frequency of the incident photons. In a gas, Raman scattering can occur with a change in vibrational, rotational or electronic energy of a molecule. Chemists are concerned primarily with the vibrational Raman Effect.

In 1922, Indian physicist C. V. Raman published his work on the Molecular Diffraction of Light, the first of a series of investigations with his collaborators which ultimately led to his discovery of the radiation effect which bears his name. The Raman Effect was first reported by C. V. Raman and K. S. Krishnan, and independently by Grigory Landsberg and Leonid Mandelstam, in 1928. Raman received the Nobel Prize in 1930 for his work on the scattering of light. In 1998 the Raman Effect was designated an ACS National Historical Chemical Landmark in recognition of its significance as a tool for analyzing the composition of liquids, gases, and solids.

Werner Heisenberg

 Werner Heisenberg, was a German theoretical physicist who made foundational contributions to quantum mechanics and is best known for asserting the uncertainty principle of quantum theory.

In addition, he made important contributions to nuclear physics, quantum field theory, and particle physics. Heisenberg, along with Max Born and Pascual Jordan, set forth the matrix formulation of quantum mechanics in 1925. Heisenberg was awarded the 1932 Nobel Prize in Physics.

Born	5 December 1901, Würzburg, Germany
Died	1 February 1976 (age 74), Munich, Germany
Nationality	German
Fields	Physics
Institutions	University of Göttingen, University of Copenhagen, University of Leipzig, University of Berlin, University of St Andrews, University of Munich
Alma mater	University of Munich
Doctoral advisor	Arnold Sommerfeld
Academic advisors	Niels Bohr, Max Born
Doctoral students	Felix Bloch, Edward Teller, Rudolph E. Peierls, Reinhard Oehme, Friedwardt Winterberg, Peter Mittelstaedt, serban Titeica Ivan Supek, Erich Bagge, Hermann Arthur Jahn, Raziuddin Siddiqui, Heimo Dolch, Hans Euler, Edwin Gora, Bernhard Kockel, Arnold Siegert, Wang Foh-san
Known for	Uncertainty Principle, Heisenberg's microscope, Matrix mechanics Kramers-Heisenberg formula, Heisenberg group, Isospin
Notable awards	Nobel Prize in Physics (1932), Max Planck Medal (1933)

UNCERTAINTY PRINCIPLE (1932)

In quantum mechanics, the Heisenberg uncertainty principle states by precise inequalities that certain pairs of physical properties, like position and momentum, cannot simultaneously be known to arbitrary precision. That is, the more precisely one property is measured, the less precisely the other can be measured. In other words, the more you know the position of a particle, the less you can know about its velocity, and the more you know about the velocity of a particle, the less you can know about its instantaneous position. According to Heisenberg its meaning is that it is impossible to determine simultaneously both the position and velocity of an electron or any other particle with any great degree of accuracy or certainty. Moreover, his principle is not a statement about the limitations of a researcher's ability to measure particular quantities of a system, but it is a statement about the nature of the system itself as described by the equations of quantum mechanics. In quantum physics, a particle is described by a wave packet, which gives rise to this phenomenon. Consider the measurement of the position of a particle. It could be anywhere the particle's wave packet has non-zero amplitude, meaning the position is uncertain–it could be almost anywhere along the wave packet. To obtain an accurate reading of position, this wave packet must be 'compressed' as much as possible, meaning it must be made up of increasing numbers of sine waves added together. The momentum of the particle is proportional to the wavelength of one of these waves, but it could be any of them. So a more accurate position measurement–by adding together more waves–means the momentum measurement becomes less accurate. The only kind of wave with a definite position is concentrated at one point, and such a wave has an indefinite wavelength. Conversely, the only kind of wave with a definite wavelength is an infinite regular periodic oscillation over all space, which has no definite position. So in quantum mechanics, there can be no states that describe a particle with both a definite position and a definite momentum. The more precise the position, the less precise the momentum.

Erwin Schrödinger

 Erwin Rudolf Josef Alexander Schrödinger, was an Austrian theoretical physicist who achieved fame for his contributions to quantum mechanics, especially the Schrödinger Equation, for which he received the Nobel Prize in Physics in 1933.

In 1935, after extensive correspondence with Albert Einstein, he proposed the Schrödinger's cat thought experiment.

Born	12 August 1887, Erdberg de, Vienna, Austria Hungary
Died	4 January 1961 (age 73), Vienna, Austria
Nationality	Austria, Citizenship: Austria, Germany, Ireland
Fields	Physics
Institutions	University of Breslau, University of Zürich, Humboldt University of Berlin, University of Oxford, University of Graz, Dublin Institute for Advanced Studies, Ghent University
Alma mater	University of Vienna
Doctoral advisor	Friedrich Hasenöhrl
Academic advisors	Franz S. Exner, Friedrich Hasenöhrl
Notable students	Linus Pauling, Felix Bloch
Known for	Schrödinger equation, Schrödinger's cat, Schrödinger method Schrödinger functional, Schrödinger picture, Schrödinger-Newton equations, Schrödinger field, Rayleigh-Schrödinger perturbation, Schrödinger logics cat state
Notable awards	Nobel Prize in Physics (1933)

SCHRÖDINGER EQUATION (1933)

In physics, specifically quantum mechanics, the Schrödinger equation is an equation that describes how the quantum state of a physical system changes in time. It is as central to quantum mechanics as Newton's laws are to classical mechanics.

In the standard interpretation of quantum mechanics, the quantum state, also called a wave function or state vector, is the most complete description that can be given to a physical system. Solutions to Schrödinger's equation describe not only molecular, atomic and subatomic systems, but also macroscopic-systems, possibly even the whole universe. The equation is named after Erwin Schrödinger, who constructed it in 1926.

The most general form is the time-dependent Schrödinger equation, which gives a description of a system evolving with time. For systems in a stationary state, the time-independent Schrödinger equation is sufficient. Approximate solutions to the time-independent Schrödinger equation are commonly used to calculate the energy levels and other properties of atoms and molecules.

Schrödinger's equation can be mathematically transformed into Werner Heisenberg's matrix mechanics, and into Richard Feynman's path integral formulation. The Schrödinger equation describes time in a way that is inconvenient for relativistic theories, a problem which is not as severe in matrix mechanics and completely absent in the path integral formulation.

The Schrödinger functional is not itself physical. It is, in its most basic form, the time translation generator of state wave functionals. In layman's terms, it defines how a system of quantum particles evolves through time and what the later systems may look like.

The basic mathematical definition is as follows. In the quantum field theory of (as example) a scalar field ϕ with a time independent Hamiltonian H the Schrödinger functional is defined as

$$S\left[\phi_2, t_2; \phi_1, t_1\right] = \left\langle \phi_2 \left| e^{-iH(t_2 - t_1)/h} \right| \phi_1 \right\rangle.$$

In the Schrödinger representation this functional generates time translations of state wave functionals

via $\Psi\left[\phi_2, t_2\right] = \int D\phi_1 S\left[\phi_2, t_2; \phi_1, t_1\right] \Psi\left[\phi_1, t_1\right].$

Paul Dirac

Paul Adrien Maurice Dirac was a British theoretical physicist. Dirac made fundamental contributions to the early development of both quantum mechanics and quantum electrodynamics.

He held the Lucasian Chair of Mathematics at the University of Cambridge and spent the last fourteen years of his life at Florida State University. Among other discoveries, he formulated the Dirac equation, which describes the behaviour of fermions and which led to the prediction of the existence of antimatter. Dirac shared the Nobel Prize in physics for 1933 with Erwin Schrödinger, for the discovery of new productive forms of atomic theory.

Born	8 August 1902, Bristol, England
Died	20 October 1984 (age 82), Tallahassee, Florida, USA
Nationality	United Kingdom
Fields	Physics
Institutions	University of Cambridge, Florida State University
Alma mater	University of Bristol, University of Cambridge
Doctoral advisor	Ralph Fowler
Doctoral students	Homi Bhabha, Harish Chandra Mehrotra, Dennis Sciama Behram Kursunoglu, John Polkinghorne
Known for	Dirac equation, Dirac comb, Dirac delta function, Fermi-Dirac statistics, Dirac sea, Dirac spinor, Dirac measure, Bracket notation Dirac adjoint, Dirac large numbers hypothesis, Dirac fermion, Dirac string, Dirac algebra, Dirac operator, Abraham-Lorentz-Dirac force, Dirac bracket, Fermi-Dirac integral, negative probability Dirac picture, Dirac-Coulomb-Breit equation
Notable awards	Nobel Prize in Physics (1933)

DIRAC EQUATION (1933)

In physics, the Dirac equation is a relativistic quantum mechanical wave equation formulated by British physicist

Paul Dirac in 1928 which provides a description of elementary spin-½ particles, such as electrons, consistent with both the principles of quantum mechanics and the theory of special relativity. The equation demands the existence of antiparticles and actually predated their experimental discovery, making the discovery of the positron, the antiparticle of the electron, one of the greatest triumphs of modern theoretical physics.

Mathematical formulation

The Dirac equation in the form originally proposed by Dirac is:

$$\left(\beta mc^2 + \sum_{k=1}^{3} \alpha_k p_k c \right) \psi(x,t) = ih \frac{\partial \psi(xt)}{\partial t}$$

where m is the rest mass of the electron, c is the speed of light,

p is the momentum operator, x and t are the space and time coordinates,

$\hbar = h/2\omega$ is the reduced Planck constant, also known as Dirac's constant.

The new elements in this equation are the 4×4 matrices α_k and β, and the four-component wave function ψ. The matrices are all Hermitian and have squares equal to the identity matrix:

$$a_i^2 = b^2 = I_4$$

and they all mutually anticommute: $\{\alpha_i, \alpha_j\} = 0$ and $\{\alpha_i, \beta\} = 0$. Explicitly

$$\alpha_i \alpha_j = -\alpha_j \alpha_i \qquad \alpha_i \beta = -\beta \alpha_i$$

where i and j are distinct and range from 1 to 3. These matrices, and the form of the wave function, have a deep mathematical significance. The algebraic structure represented by the Dirac matrices had been created some 50 years earlier by the English mathematician W. K. Clifford, which in turn had been based on the mid-19th century work of the German mathematician Hermann Grassmann in his Lineare Ausdehnungslehre. The latter had been regarded as well-nigh incomprehensible by most of his contemporaries. The appearance of something so seemingly abstract, at such a late date, in such a direct physical manner, amounts to one of the most remarkable chapters in the history of physics.

The commutation rules are designed so that a solution of Dirac's equation will automatically also be a solution of

$$\left((mc^2)^2 + \sum_{k=1}^{3} (p_k c)^2 \right) \psi = E^2 \psi$$

this is the relativistic energy-momentum equation.

James Chadwick

Sir James Chadwick, was an English Nobel laureate in physics awarded for his discovery of the neutron. In 1932, Chadwick discovered a previously unknown particle in the atomic nucleus. This particle became known as the neutron because of its lack of electric charge. Chadwick's discovery was crucial for the fission of uranium 235.

Unlike positively charged alpha particles, which are repelled by the electrical forces present in the nuclei of other atoms, neutrons do not need to overcome any Coulomb barrier and can therefore penetrate and split the nuclei of even the heaviest elements. For this discovery he was awarded the Hughes Medal of the Royal Society in 1932 and the Nobel Prize for Physics in 1935.

Born	20 October 1891, Bollington, Cheshire, England
Died	24 July 1974 (age 82), Cambridge, England
Nationality	Citizenship United Kingdom
Fields	Physics
Institutions	Technical University of Berlin, Liverpool University, Gonville and Caius College, Cambridge University, Manhattan Project.
Alma mater	University of Manchester, University of Cambridge
Academic advisor	Ernest Rutherford, Hans Geiger
Doctoral students	Maurice Goldhaber, Ernest C. Pollard, Charles Drummond Ellis Dai Chuanzeng
Known for	Discovery of the neutron
Notable awards	Nobel Prize in Physics (1935), Franklin Medal (1951)

NEUTRON (1935)

The neutron is a subatomic particle with no net electric charge and a mass slightly larger than that of a proton. They are usually found in atomic nuclei. The nuclei of most atoms consist of protons and neutrons, which are therefore collectively referred to as nucleons. The number of protons in a nucleus is the atomic number and defines the type of element the atom forms. The number of neutrons is the neutron number and determines the isotope of an element. While bound neutrons in stable nuclei are stable, free neutrons are unstable; they undergo beta decay with a mean lifetime of just under 15 minutes ($885.7 \pm 0.8s$). Free neutrons are produced in nuclear fission and fusion. Dedicated neutron sources like research reactors and spallation sources produce free neutrons for use in irradiation and in neutron scattering experiments. Even though it is not a chemical element, the free neutron is sometimes included in tables of nuclides. It is then considered to have an atomic number of zero and a mass number of one, and is sometimes referred to as neutronium

In 1931 Walther Bothe and Herbert Becker in Germany found that if the very energetic alpha particles emitted from polonium fell on certain light elements, specifically beryllium, boron, or lithium, an unusually penetrating radiation was produced. At first this radiation was thought to be gamma radiation, although it was more penetrating than any gamma rays known, and the details of experimental results were very difficult to interpret on this basis. The next important contribution was reported in 1932 by Irène Joliot-Curie and Frederic Joliot in Paris. They showed that if this unknown radiation fell on paraffin, or any other hydrogen-containing compound, it ejected protons of very high energy. This was not in itself inconsistent with the assumed gamma ray nature of the new radiation, but detailed quantitative analysis of the data became increasingly difficult to reconcile with such a hypothesis. In 1932, James Chadwick performed a series of experiments at the University of Manchester, showing that the gamma ray hypothesis was untenable. He suggested that the new radiation consisted of uncharged particles of approximately the mass of the proton, and he performed a series of experiments verifying his suggestion. These uncharged particles were called *neutrons*, apparently from the Latin root for *neutral* and the Greek ending *-on*.

Victor Francis Hess

Victor Francis Hess was an Austrian-American physicist, and Nobel laureate in physics, who discovered cosmic rays. Hess undertook the work that won him the Nobel Prize in Physics in 1936.

Born	24 June 1883, Schloss Waldstein, Peggau, Austria
Died	17 Dec 1964 (age 81), Mount Vernon, New York, USA
Nationality	Austria, United States
Fields	Physics
Institutions	University of Graz, Austrian Academy of Sciences, University of Innsbruck, Fordham University
Alma mater	University of Graz
Known for	Discovery of cosmic rays
Notable awards	Nobel Prize in Physics (1936)

COSMIC RAY (1936)

Cosmic rays are energetic particles originating from outer space that impinge on Earth's atmosphere. Almost 90% of all the incoming cosmic ray particles are simple protons, with nearly 10% being helium nuclei (alpha particles), and slightly under 1% are heavier elements, electrons (beta particles), or gamma ray photons. The term *ray* is a misnomer, as cosmic particles arrive individually, not in the form of a ray or beam of particles. However, when they were first discovered, cosmic rays were thought to be rays. When their particle nature needs to be emphasized, cosmic ray particle is written.

The variety of particle energies reflects the wide variety of sources. The origins of these particles range from energetic processes on the Sun all the way to as yet unknown events in the farthest reaches of the visible universe. Cosmic rays can have energies of over 10^{20} eV, far higher than the 10^{12} to 10^{13} eV that man-made particle accelerators can produce. Ultra-high-energy cosmic rays for a description of the detection of a single particle with an energy of about 50 J, the same as a well-hit tennis ball at 42 m/s [about 150 km/h]. There has been interest in investigating cosmic rays of even greater energies.

The nucleus that make up cosmic rays are able to travel from their distant sources to the Earth because of the low density of matter in space. Nuclei interact strongly with other matter, so when the cosmic rays approach Earth they begin to collide with the nuclei of atmospheric gases. These collisions, in a process known as a shower, result in the production of many pions and kaons, unstable mesons which quickly decay into muons. Because muons do not interact strongly with the atmosphere and because of the relativistic effect of time dilation many of these muons are able to reach the surface of the Earth. Muons are ionizing radiation, and may easily be detected by many types of particle detectors such as bubble chambers or scintillation detectors. If several muons are observed by separated detectors at the same instant it is clear that they must have been produced in the same shower event.

Cosmic rays impacting other (non-Earth) bodies in the solar system which are made of elements heavier than hydrogen and helium, can be detected indirectly by observing high energy gamma ray emissions from these bodies using a gamma-ray telescope. When such gammas are of energy too high to result from radioactive decay processes (> about 10 Mev) they must be secondary to cosmic ray bombardment.

The nuclei that make up cosmic rays are able to travel from their distant sources to the Earth because of the low density of matter in space. Nuclei interact strongly with other matter, so when the cosmic rays approach Earth they begin to collide with the nuclei of atmospheric gases. These collisions, in a process known as a shower, result in the production of many pions and kaons, unstable mesons which quickly decay into muons.

Carl David Anderson

 Carl David Anderson was an American physicist. He is best known for his discovery of the positron in 1932, an achievement for which he received the 1936 Nobel Prize in Physics, and of the muon in 1936.

Born	3 September 1905, New York City, New York, USA
Died	11 January 1991 (age 85), San Marino, California, USA
Nationality	United States
Fields	Physicist
Institutions	California Institute of Technology
Alma mater	California Institute of Technology
Notable students	Donald A. Glaser, Seth Neddermeyer
Known for	Discovery of the positron, discovery of the muon
Notable awards	Nobel Prize in Physics (1936), Elliott Cresson Medal (1937)

POSITRON (1936)

The positron or antielectron is the antiparticle or the antimatter counterpart of the electron. The positron has an electric charge of +1e, a spin of $\frac{1}{2}$, and the same mass as an electron. When a low-energy positron collides with a low-energy electron, annihilation occurs, resulting in the production of two or more gamma ray photons. Positrons may be generated by positron emission radioactive decay, or by pair production from a sufficiently energetic photon.

Dmitri Skobeltsyn came close to discovering the positron in 1923. While using a bubble chamber to try to detect gamma radiation in cosmic rays, Skobeltsyn detected particles that acted like electrons but curved in the opposite direction in an applied magnetic field. He was puzzled by these results, and shared them with other scientists, none of whom could explain them. Likewise, in 1929 Chung-Yao Chao, a graduate student at Caltech, noticed some anomalous results that indicated particles behaving like electrons, but with a positive charge, though the results were inconclusive and the phenomenon was not pursued.

Anderson also coined the term *positron*. The positron was the first evidence of antimatter and was discovered when Anderson allowed cosmic rays to pass through a cloud chamber and a lead plate. A magnet surrounded this apparatus, causing particles to bend in different directions based on their electric charge. The ion trail left by each positron appeared on the photographic plate with a curvature matching the mass-to-charge ratio of an electron, but in a direction that proved its charge was positive. Anderson wrote in retrospect that the positron could have been discovered earlier based on Chung-Yao Chao's work, if only it had been followed up.

The muon is an elementary particle similar to the electron, with a negative electric charge and a spin of $\frac{1}{2}$. Together with the electron, the tauon, and the three neutrinos, it is classified as a lepton. It is an unstable subatomic particle with the second longest mean lifetime (2.2 µs), exceeded only by that of the free neutron (~15 min). Like all elementary particles, the muon has a corresponding antiparticle of opposite charge but equal mass and spin: the antimuon. Muons are denoted by μ^- and antimuons by μ^+. Muons were sometimes referred to as mu mesons in the past, even though they are not classified as mesons by modern particle physicists.

Muons have a mass of 105.7 Mev/c^2, which is about 200 times the mass of the electrons. Since the muon's interactions are very similar to those of the electron, a muon can be thought of in most ways as simply a much heavier version of the electron.

Clinton Davisson

Clinton Joseph Davisson was an American physicist who won the 1937 Nobel Prize in Physics for his discovery of electron diffraction. Davisson shared the Nobel Prize with George Paget Thomson, who independently discovered electron diffraction at about the same time as Davisson.

Born	22 October 1881, Bloomington, Illinois, USA
Died	1 February 1958 (age 76), Charlottesville, Virginia, USA
Nationality	United States
Fields	Physics
Institutions	Princeton University, Carnegie Institute of Technology, Bell Labs
Alma mater	University of Chicago, Princeton University
Doctoral advisor	Owen Richardson
Known for	Electron diffraction
Notable awards	Comstock Prize (1928), Elliott Cresson Medal (1931), Nobel Prize in Physics (1937)

ELECTRON DIFFRACTION (1937)

Electron diffraction refers to the wave nature of electrons. However, from a technical or practical point of view, it may be regarded as a technique used to study matter by firing electrons at a sample and observing the resulting interference pattern. This phenomenon is commonly known as the wave-particle duality, which states that the *behavior* of a particle of matter (in this case the incident electron) can be described by a wave. For this reason, an electron can be regarded as a wave much like sound or water waves. This technique is similar to X-ray and neutron diffraction.

Electron diffraction is most frequently used in solid state physics and chemistry to study the crystal structure of solids. Experiments are usually performed in a transmission electron microscope (TEM), or a scanning electron microscope (SEM) as electron backscatter diffraction. In these instruments, electrons are accelerated by an electrostatic potential in order to gain the desired energy and determine their wavelength before they interact with the sample to be studied.

The periodic structure of a crystalline solid acts as a diffraction grating, scattering the electrons in a predictable manner. Working back from the observed diffraction pattern, it may be possible to deduce the structure of the crystal producing the diffraction pattern. However, the technique is limited by the phase problem. Apart from the study of crystals, i.e. electron crystallography, electron diffraction is also a useful technique to study the short range order of amorphous solids, and the geometry of gaseous molecules.

Diffraction is a characteristic effect when a wave is incident upon an aperture or a grating, and is closely associated with the meaning of wave motion itself. In the 19th Century, diffraction was well-established for light and for ripples on the surfaces of fluids. In 1927, while working for Bell Labs, Davisson and Lester Germer performed an experiment showing that electrons were diffracted at the surface of a crystal of nickel. This celebrated Davisson-Germer experiment confirmed the de Broglie hypothesis that particles of matter have a wave-like nature, which is a central tenet of quantum mechanics. In particular, their observation of diffraction allowed the first measurement of a wavelength for electrons. The measured wavelength λ agreed well with de Broglie's equation $\lambda = h / p$, where h is Planck's constant and p is the electron's momentum.

George Paget Thomson

 Sir George Paget Thomson was an English physicist and Nobel laureate in physics recognized for his discovery with Clinton Davisson of the wave properties of the electron by electron diffraction. Thomson became a Fellow at Cambridge and then moved to the University of Aberdeen.

George Thomson was jointly awarded the Nobel Prize for Physics in 1937 for his work in Aberdeen in discovering the wave-like properties of the electron. The prize was shared with Clinton Joseph Davisson who had made the same discovery independently. Whereas his father had seen the electron as a particle, Thomson demonstrated that it could be diffracted like a wave, a discovery proving the principle of wave-particle duality which had first been posited by Louis-Victor de Broglie in the 1920s as what is often dubbed the de Broglie hypothesis.

Born	3 May 1892, Cambridge, England
Died	10 September 1975 (age 83), Cambridge, England
Nationality	United Kingdom
Fields	Physics
Institutions	University of Aberdeen, University of Cambridge, Imperial College London
Alma mater	University of Cambridge
Doctoral advisor	John Strutt (Rayleigh)
Doctoral students	Ishrat Hussain Usmani
Known for	Electron diffraction
Notable awards	Howard N. Potts Medal (1932), Nobel Prize in Physics (1937)

ELECTRON DIFFRACTION (1937)

The de Broglie hypothesis, formulated in 1926, predicts that particles should also behave as waves. De Broglie's formula was confirmed three years later for electrons with the observation of electron diffraction in two independent experiments. At the University of Aberdeen George Paget Thomson passed a beam of electrons through a thin metal film and observed the predicted interference patterns. At Bell Labs Clinton Joseph Davisson and Lester Halbert Germer guided their beam through a crystalline grid. Thomson and Davisson shared the Nobel Prize for Physics in 1937 for their work.

ELECTRON INTERACTION WITH MATTER

Unlike other types of radiation used in diffraction studies of materials, such as X-rays and neutrons, electrons are charged particles and interact with matter through the Coulomb forces. This means that the incident electrons feel the influence of both the positively charged atomic nuclei and the surrounding electrons. In comparison, X-rays interact with the spatial distribution of the valence electrons, while neutrons are scattered by the atomic nuclei through the strong nuclear forces. In addition, the magnetic moment of neutrons is non-zero, and they are therefore also scattered by magnetic fields. Because of these different forms of interaction, the three types of radiation are suitable for different studies.

INTENSITY OF DIFFRACTED BEAMS

In the kinematical approximation for electron diffraction, the intensity of a diffracted beam is given by:

$$I_g = |\psi_g|^2 \, \alpha |F_g|^2$$

here Ψ_g is the wave function of the diffracted beam and F_g is the so called structure factor which is given by:

$$F_g = \sum_i f_i e^{-2\pi i g \cdot r_i}$$

where g is the scattering vector of the diffracted beam, r_1 is the position of an atom i in the unit cell, and f_i is the scattering power of the atom, also called the atomic form factor. The sum is over all atoms in the unit cell.

Enrico Fermi

Enrico Fermi was an Italian physicist, particularly remembered for his work on the development of the first nuclear reactor, and for his contributions to the development of quantum theory, nuclear and particle physics, and statistical mechanics.

Awarded the Nobel Prize in Physics in 1938 for his work on induced radioactivity. Fermi is widely regarded as one of the leading scientists of the 20th century, highly accomplished in both theory and experiment. Fermium, a synthetic element created in 1952.

Born	29 September 1901, Rome, Italy
Died	28 November 1954 (age 53), Chicago, Illinois, USA
Nationality	Citizenship Italy (1901–1954), United States (1944–1954)
Fields	Physicist
Institutions	Scuola Normale Superiore in Pisa, University of Göttingen, University of Leiden, University of Rome La Sapienza, Columbia University, University of Chicago
Alma mater	Scuola Normale Superiore
Doctoral advisor	Luigi Puccianti
Doctoral students	Owen Chamberlain, Geoffrey Chew, Mildred Dresselhaus, Jerome I. Friedman, Marvin Leonard Goldberger, Tsung-Dao Lee, Ettore Majorana, James Rainwater, Marshall Rosenbluth, Arthur Osenfeld, Emilio Segrè, Jack Steinberger, Sam Treiman
Known for	New radioactive elements produced by neutron irradiation, controlled nuclear chain reaction, Fermi-Dirac statistics theory of beta decay
Notable awards	Matteucci Medal (1926), Nobel Prize in Physics (1938), Hughes Medal (1942), Franklin Medal (1947), Rumford Prize (1953)

INDUCED RADIOACTIVITY (1938)

Induced radioactivity occurs when a previously stable material has been made radioactive by exposure to specific radiation. Most radioactivities do not induce other material to become radioactive.

Neutron activation is the main form of induced radioactivity, which happens when free neutrons are captured by nuclei. This new heavier isotope can be stable or unstable depending on the chemical element involved. Because free neutrons disintegrate within minutes outside of an atomic nucleus, neutron radiation can be obtained only from nuclear disintegrations, nuclear reactions, and high-energy reactions. Neutrons that have been slowed down through a neutron moderator are more likely to be captured by nuclei than fast neutrons.

A less common form involves removing a neutron via photodisintegration. In this reaction, a high energy photon strikes a nucleus with an energy greater than the binding energy of the atom, releasing a neutron. This reaction has a minimum cutoff of 2 Mev and around 10 Mev for most heavy nuclei. Many radionuclides do not produce gamma rays with energy high enough to induce this reaction. The isotopes used in food irradiation both have energy peaks below this cutoff and thus cannot induce radioactivity in the food.

Some induced radioactivity is produced by background radiation, which is mostly natural. However, since natural radiation is not very intense in most places on Earth, the amount of induced radioactivity in a single location is usually very small.

The conditions inside certain types of nuclear reactors with high neutron flux can cause induced radioactivity. The components in those reactors may become highly radioactive from the radiation to which they are exposed. Induced radioactivity increases the amount of nuclear waste that must eventually be disposed, but it is not referred to as radioactive contamination unless it is uncontrolled.

Ernest Lawrence

 Ernest Orlando Lawrence was an American physicist and Nobel laureate, known for his invention, utilization, and improvement of the cyclotron atom-smasher beginning in 1929, and his later work in uranium-isotope separation for the Manhattan Project.

Lawrence had a long career at the University of California, where he became a Professor of Physics. In 1939, Lawrence was awarded the Nobel Prize in Physics for his work in inventing the cyclotron and developing its applications. Chemical element number 103 is named lawrencium in Lawrence's honor. He was also the first recipient of the Sylvanus Thayer Award.

Born	8 August 1901, Canton, South Dakota
Died	27 August 1958 (age 57), Palo Alto, California
Nationality	American
Residence	United States
Fields	Physics
Institutions	Yale University, University of California
Alma mater	University of South Dakota, University of Minnesota, Yale University
Doctoral advisor	W.F.G. Swann
Doctoral students	Edwin McMillan, Chien-Shiung Wu
Known for	The invention of the cyclotron atom-smasher, elementary particle physics, The Manhattan Project
Notable awards	Elliott Cresson Medal (1937), Nobel Prize in Physics (1939) Enrico Fermi Award (1957)

CYCLOTRON (1939)

In technology, a cyclotron is a type of particle accelerator. In physics, cyclotron frequency is the frequency of a charged particle moving perpendicular to the direction of a uniform magnetic field, i.e. a magnetic field of constant magnitude and direction. Since that motion is always circular, the cyclotron frequency is well defined.

Cyclotrons accelerate charged particles using a high-frequency, alternating voltage. A perpendicular magnetic field causes the particles to spiral almost in a circle so that they re-encounter the accelerating voltage many times.

The first cyclotron was manufactured by Ernest Lawrence, of the University of California, Berkeley who started operating it in 1932, though others had been working along similar lines at the time. The first European cyclotron was founded in Leningrad in the physics department of the Radium Institute. In 1932 George Gamow and Lev Mysovskii presented a draft for consideration by the Scientific Council of the Radium Institute, and the approval of it, under the guidance and direct participation of the Igor Kurchatov and Lev Mysovskii cyclotron was installed and running by 1937.

TRIUMF, Canada's national laboratory for nuclear and particle physics, houses the world's largest cyclotron. The 18m diameter, 4000 tonne main magnet produces a field of 0.46 T while a 23 MHz 94 kV electric field is used to accelerate the 200 µA beams. TRIUMF conducts world class science and is located at the University of British Columbia, Vancouver, Canada, and is run as a consortium of fifteen Canadian universities.

For several decades, cyclotrons were the best source of high-energy beams for nuclear physics experiments; several cyclotrons are still in use for this type of research.

Cyclotrons can be used to treat cancer. Ion beams from cyclotrons can be used, as in proton therapy, to penetrate the body and kill tumors by radiation damage, while minimizing damage to healthy tissue along their path.

Cyclotron beams can be used to bombard other atoms to produce short-lived positron-emitting isotopes suitable for PET imaging.

Otto Stern

Otto Stern was a German physicist and Nobel laureate in physics. He was awarded the 1943 Nobel Prize in Physics, the first to be awarded since 1939. He was the sole recipient in Physics that year, and the award citation omitted mention of the Stern-Gerlach experiment, as Gerlach had remained active in Nazi-led Germany.

Born	17 February 1888, Sohrau, Kingdom of Prussia
Died	17 August 1969 (age 81), Berkeley, California, USA
Nationality	Germany
Fields	Physics
Institutions	University of Rostock, University of Hamburg, Carnegie Institute of Technology, University of California, Berkeley
Alma mater	University of Breslau, University of Frankfurt
Known for	Stern-Gerlach experiment, Spin quantization, Molecular ray method
Notable awards	Nobel Prize in Physics (1943)

STERN-GERLACH EXPERIMENT (1943)

In quantum mechanics, the Stern-Gerlach experiment, named after Otto Stern and Walther Gerlach, is an important 1922 experiment on the deflection of particles, often used to illustrate basic principles of quantum mechanics. It can be used to demonstrate that electrons and atoms have intrinsically quantum properties, and how measurement in quantum mechanics affects the system being measured.

The Stern-Gerlach experiment was performed in Frankfurt, Germany in 1922 by Otto Stern and Walther Gerlach. At the time, Stern was an assistant to Max Born at the University of Frankfurt's Institute for Theoretical Physics, and Gerlach was an assistant at the same university's Institute for Experimental Physics.

At the time of the experiment, the most prevalent model for describing the atom was the Bohr model, which described electrons as going around the positively-charged nucleus only in certain discrete atomic orbital's or energy levels. Since the electron was quantized to be only in certain positions in space, the separation into distinct orbits was referred to as space quantization. The Stern-Gerlach experiment was meant to test the Bohr-Sommerfeld hypothesis that the direction of the angular momentum of a silver atom is quantized.

Note that the experiment was performed several years before Uhlenbeck and Goudsmit formulated their hypothesis of the existence of the electron spin. Even though the result of the Stern-Gerlach experiment has later turned out to be in agreement with the predictions of quantum mechanics for a spin-$\frac{1}{2}$ particle, the experiment should be seen as a corroboration of the Bohr-Sommerfeld theory.

In 1927, T.E. Phipps and J.B. Taylor reproduced the effect using hydrogen atoms in their ground state, thereby eliminating any doubts that may have been caused by the use of silver atoms. In 1926 the non-relativistic Schrödinger equation had incorrectly predicted the magnetic moment of hydrogen to be zero in its ground state. To correct this problem Pauli introduced by hand so to speak, the 3 spin matrices which now bear his name, but which were then later shown by Dirac in 1928 to be intrinsic in his relativistic equation.

The Stern-Gerlach experiment had one of the biggest impacts on modern physics:

- In the decade that followed, scientists showed using similar techniques, that the nuclei of some atoms also have quantized angular momentum. It is the interaction of this nuclear angular momentum with the spin of the electron that is responsible for the hyperfine structure of the spectroscopic lines.

Isidor Isaac Rabi

Isidor Isaac Rabi was a Galician born American physicist and Nobel laureate recognized in 1944 for his discovery of nuclear magnetic resonance. In 1930 Rabi conducted investigations into the nature of the force binding protons to atomic nuclei. This research eventually led to the creation of the molecular-beam magnetic-resonance detection method.

Born	29 July 1898, Rymanów, Galicia, Austria-Hungary
Died	11 Jan 1988 (aged 89), New York City, New York, USA
Nationality	United States
Fields	Physics
Institutions	Columbia University, MIT
Alma mater	Cornell University, Columbia University
Doctoral advisor	Albert Potter Wills
Doctoral students	Julian Schwinger, Norman F. Ramsey, Martin L. Perl
Known for	Nuclear magnetic resonance
Notable awards	Elliott Cresson Medal (1942), Nobel Prize in Physics (1944)

NUCLEAR MAGNETIC RESONANCE (1944)

Nuclear magnetic resonance (NMR) is a property that magnetic nuclei have in a magnetic field and applied electromagnetic (EM) pulse or pulses, which cause the nuclei to absorb energy from the EM pulse and radiate this energy back out. The energy radiated back out is at a specific resonance frequency which depends on the strength of the magnetic field and other factors. This allows the observation of specific quantum mechanical magnetic properties of an atomic nucleus. Many scientific techniques exploit NMR phenomena to study molecular physics, crystals and non-crystalline materials through NMR spectroscopy. NMR is also routinely used in advanced medical imaging techniques, such as in magnetic resonance imaging (MRI).

All stable isotopes that contain an odd number of protons and/or of neutrons have an intrinsic magnetic moment and angular momentum, in other words a nonzero spin, while all nuclides with even numbers of both have spin 0. The most commonly studied nuclei are ^1H (the most NMR-sensitive isotope after the radioactive ^3H) and ^{13}C, although nuclei from isotopes of many other elements (e.g. ^2H, ^{10}B, ^{11}B, ^{14}N, ^{15}N, ^{17}O, ^{19}F, ^{23}Na, ^{29}Si, ^{31}P, ^{35}Cl, ^{113}Cd, ^{129}Xe, ^{195}Pt) are studied by high-field NMR spectroscopy as well.

A key feature of NMR is that the resonance frequency of a particular substance is directly proportional to the strength of the applied magnetic field. It is this feature that is exploited in imaging techniques; if a sample is placed in a non-uniform magnetic field then the resonance frequencies of the sample's nuclei depend on where in the field they are located. Since the resolution of the imaging techniques depends on how big the gradient of the field is, many efforts are made to develop more powerful magnets, often using superconductors. The effectiveness of NMR can also be improved using hyper polarization, and/or using two-dimensional, three-dimensional and higher dimension multi-frequency techniques.

The principle of NMR usually involves two sequential steps:

- The alignment (polarization) of the magnetic nuclear spins in an applied, constant magnetic field H_0.
- The perturbation of this alignment of the nuclear spins by employing an electromagnetic, usually radio frequency (RF) pulse. The required perturbing frequency is dependent upon the static magnetic field (H_0) and the nuclei of observation.

Wolfgang Pauli

 Wolfgang Ernst Pauli was an Austrian theoretical physicist and one of the pioneers of quantum physics. In 1945, after being nominated by Albert Einstein, he received the Nobel Prize in Physics for his decisive contribution through his discovery of a new law of Nature, the exclusion principle or Pauli principle, involving spin theory, underpinning the structure of matter and the whole of chemistry.

Born	25 April 1900, Vienna, Austria-Hungary
Died	15 December 1958 (age 58), Zürich, Switzerland
Nationality	Austria
Citizenship	Switzerland
Fields	Physics
Institutions	University of Göttingen, University of Copenhagen, University of Hamburg, ETH Zürich, Princeton University
Alma mater	Ludwig-Maximilians University
Doctoral advisor	Arnold Sommerfeld
Academic advisors	Max Born
Doctoral students	Nicholas Kemmer, Felix Villars, Sigurd Zienau
Known for	Pauli exclusion principle, Pauli-Villars regularization, Pauli matrices, Pauli effect, Pauli equation, Pauli group, Coining 'not even wrong'
Notable awards	Lorentz Medal (1931), Nobel Prize in Physics (1945), Matteucci Medal (1956), Max Planck Medal (1958)

PAULI EXCLUSION PRINCIPLE (1945)

The Pauli Exclusion Principle is a quantum mechanical principle formulated by the Austrian physicist Wolfgang Pauli in 1925. In its simplest form for electrons in a single atom, it states that no two electrons can have the same four quantum numbers, that is, if n, l, and m_l are the same, m_s must be different such that the electrons have opposite spins. More generally, no two identical fermions may occupy the same quantum state simultaneously. A more rigorous statement of this principle is that, for two identical fermions, the total wave function is anti-symmetric. In contrast, integer spin particles, bosons are not subject to the Pauli exclusion principle. For bosons any number of identical particles can occupy the same quantum state, this is seen in for instance lasers and Bose-Einstein condensation. In the early 20th century, it became evident that atoms and molecules with even numbers of electrons are more stable than those with odd numbers of electrons. In the famous 1916 article *The Atom and the Molecule* by Gilbert N. Lewis, for example, rule three of his six postulates of chemical behavior states that the atom tends to hold an even number of electrons in the shell and especially to hold eight electrons which are normally arranged symmetrically at the eight corners of a cube.

In 1919, Irving Langmuir suggested that the periodic table could be explained if the electrons in an atom were connected or clustered in some manner. Groups of electrons were thought to occupy a set of electron shells about the nucleus. In 1922, Niels Bohr updated his model of the atom by assuming that certain numbers of electrons corresponded to stable closed shells. Pauli looked for an explanation for these numbers, which were at first only empirical. At the same time he was trying to explain experimental results in the Zeeman Effect in atomic spectroscopy and in ferromagnetism. He found an essential clue in 1924 paper by E.C. Stoner which pointed out that for a given value of the principal quantum number (n), the number of energy levels of a single electron in the alkali metal spectra in an external magnetic field, where all degenerate energy levels are separated, is equal to the number of electrons in the closed shell of the rare gases for the same value of n. This led Pauli to realize that the complicated numbers of electrons in closed shells can be reduced to the simple rule of *one* per state, if the electron states are defined using four quantum numbers. For this purpose he introduced a new two-valued quantum number, identified by Samuel Goudsmit and George Uhlenbeck as electron spin.

Percy Williams Bridgman

 Percy Williams Bridgman was an American physicist who won the 1946 Nobel Prize in Physics for his work on the physics of high pressures. He also wrote extensively on the scientific method and on other aspects of the philosophy of science.

Born	21 April 1882, Cambridge, Massachusetts, USA
Died	20 August 1961 (age 79), Randolph, New Hampshire, USA
Nationality	United States
Fields	Physics
Institutions	Harvard University
Alma mater	Harvard University
Doctoral advisor	Wallace Clement Sabine
Doctoral students	John C. Slater, John Hasbrouck Van Vleck
Known for	High pressure physics
Notable awards	Rumford Prize (1917), Elliott Cresson Medal (1932), Nobel Prize in Physics (1946)

He began investigating the properties of matter under high pressure. A machinery malfunction led him to modify his pressure apparatus; the result was a new device enabling him to create pressures eventually exceeding 100,000 kgf/cm² (10 GPa). This was a huge improvement over previous machinery, which could achieve pressures of only 3,000 kgf/cm² (0.3 GPa). This new apparatus led to an abundance of new findings, including on the effect of pressure on electrical resistance, and on the liquid and solid states. Bridgman is also known for his studies of electrical conduction in metals and properties of crystals. He developed the Bridgman seal and is the eponym for Bridgman's thermodynamic equations.

Bridgman made many improvements to his high pressure apparatus over the years, and unsuccessfully attempted the synthesis of diamond many times. His writings on the philosophy of science advocated operationalism, and he coined the term operational definition. He was also one of the 11 signatories to the Russell-Einstein Manifesto.

HIGH PRESSURE PHYSICS (1946)

High pressure science and engineering is studying the effects of high pressure on materials and the design and construction of devices, such as a diamond anvil cell, which can create high pressure. By high pressure it is usually meant pressures of thousands (kilobars) or millions (megabars) of times atmospheric pressure (about 1 bar).

It was by applying high pressure as well as high temperature to carbon that man-made diamonds were first produced as well as many other interesting discoveries. Almost any material when subjected to high pressure will compact itself into a denser form, for example, quartz, also called silica or silicon dioxide will first adopt a denser form known as coesite, then upon application of more temperature, form stishovite. These two forms of silica were first discovered by high pressure experimenters, but then found in nature at the site of a meteor impact.

Chemical bonding is likely to change under high pressure, when the P*V term in the free energy becomes comparable to the energies of typical chemical bonds, i.e. at around 100 GPa. Among the most striking changes are metallization of oxygen at 96 GPa (rendering oxygen a superconductor), and transition of sodium from a nearly free electron metal to a transparent insulator at –200 GPa. At ultimately high compression, however, all materials will metalize.

High pressure experimentation has led to the discovery of the types of minerals which are believed to exist in the deep mantle of the Earth, such as perovskite which is thought to make up half of the Earth's bulk, and post-perovskite, which occurs at the core-mantle boundary and explains many anomalies inferred for that region.

Edward Victor Appleton

Sir Edward Victor Appleton, was an English physicist. He received the 1947 Nobel Prize in Physics for his contributions to the knowledge of the ionosphere, which led to the development of radar.

Born	6 September 1892, Bradford, West Yorkshire, England
Died	21 April 1965 (age 72), Edinburgh, Scotland
Nationality	English
Fields	Physics
Institutions	Cambridge University, King's College London, Edinburgh University
Alma mater	Cambridge University
Academic advisor	J. J. Thomson, Ernest Rutherford
Notable students	J. A. Ratcliffe, Charles Oatley
Known for	Ionospheric Physics, Appleton layer
Notable awards	Nobel Prize in Physics (1947)

IONOSPHERE (1947)

The ionosphere is the uppermost part of the atmosphere, between the thermosphere and the exosphere, distinguished because it is ionized by solar radiation. It plays an important part in atmospheric electricity and forms the inner edge of the magnetosphere. It has practical importance because, among other functions, it influences radio propagation to distant places on the Earth. The F region of the ionosphere is home to the F layer of ionization, also called the Appleton layer, after the English physicist Edward Appleton. As with other ionospheric sectors, layer implies a concentration of plasma, while region is the area that contains the said layer. The F region contains ionized gases at a height of around 150–800 km above sea level, placing it in the Earth's thermosphere, a hot region in the upper atmosphere, and also in the heterosphere, where chemical composition varies with height. Generally speaking, the F region has the highest concentration of free electrons and ions anywhere in the atmosphere. It may be thought of as comprising two layers, the F_1-and F_2-layers.

The F-region is located directly above the E region and below the protonosphere. It acts as a dependable reflector of radio signals as it is not affected by atmospheric conditions, although its ionic composition varies with the sunspot cycle. It reflects normal-incident frequencies at or below the critical frequency (~ 10 MHz) and partially absorbs waves of higher frequency. The F region is the region of the ionosphere which is very important for HF radio wave propagation. This F region is very anomalous in nature. The F_1 layer is the lower sector of the F layer and exists from about 150 to 220 km above the surface of the Earth and only during daylight hours. It is composed of a mixture of molecular ions O_2^+ and NO^+, and atomic ions O^+. Above the F_1 region, atomic oxygen becomes the dominant constituent because lighter particles tend to occupy higher altitudes above the turbo pause (at ~100 km). This atomic oxygen provides the O^+ atomic ions that make up the F_2 layer. The F_1 layer has approximately 5×10^5 e/cm^3 at noontime and minimum sunspot activity, and increases to roughly 2×10^6 e/cm^3 during maximum sunspot activity. The density falls off to below 10^4 e/cm^3 at night.

- The F_1 layer merges into the F_2 layer at night.
- Though fairly regular in its characteristics, it is not observable everywhere or on all days. The principal reflecting layer during the summer for paths of 2,000 to 3,500 km is the F_1 layer.
- The F_2 layer exists from about 220 to 800 km above the surface of the Earth. The F_2 layer is the principal reflecting layer for HF communications during both day and night.

Patrick Blackett, Baron Blackett

 Patrick Maynard Stuart Blackett, Baron Blackett was an English experimental physicist known for his work on cloud chambers, cosmic rays, and paleomagnetism. He also made a major contribution in World War II advising on military strategy and developing Operational Research.

His left-wing views saw an outlet in third world development and in influencing policy in the Labour Government of the 1960s. In 1948 he was awarded the Nobel Prize in Physics, for his investigation of cosmic rays using his invention of the counter-controlled cloud chamber.

Born	18 November 1897, London, England
Died	13 July 1974 (age 76), London, England
Nationality	United Kingdom
Fields	Physics
Institutions	University of Cambridge, University of London, University of Manchester, Imperial College
Alma mater	Osborne Naval College, Cambridge University
Doctoral advisor	Ernest Rutherford
Doctoral students	Edward Bullard, Ishrat Hussain Usmani
Known for	Cloud chambers, Cosmic rays, Paleomagnetism
Notable awards	Nobel Prize in Physics (1948)

PALEOMAGNETISM (1948)

Paleomagnetism is the study of the record of the Earth's magnetic field preserved in various magnetic minerals through time. The study of paleomagnetism has demonstrated that the Earth's magnetic field varies substantially in both orientation and intensity through time.

A paleomagnetist is a scientist who studies the ancient magnetic field by measuring the magnetic direction recorded in minerals in rocks and sediments, acquired at the time of their formation (remanent magnetization), then using methods similar to geomagnetism to determine what configuration of the Earth's magnetic field may have resulted in the observed orientation.

Paleomagnetism is studied on a number of scales:

- *Secular variation studies* look at small-scale changes in the direction and intensity of the Earth's magnetic field. The magnetic north pole is constantly shifting relative to the axis of rotation of the Earth. Magnetism is a vector and so magnetic field variation is made up of palaeodirectional measurements of magnetic declination and magnetic inclination and palaeointensity measurements.
- *Reversal magnetostratigraphy* examines the periodic polarity reversion of the Earth's magnetic field. The reversals have occurred at irregular intervals throughout the Earth's history. The age and pattern of these reversals is known from the study of sea floor spreading zones and the dating of volcanic rocks.

The study of paleomagnetism is possible because iron-bearing minerals such as magnetite may record past directions of the Earth's magnetic field. Paleomagnetic signatures in rocks can be recorded by three different mechanisms namely; Thermal remanent magnetization, Detrital remanent magnetization and Chemical remanent magnetization.

Paleomagnetic evidence, both reversals and polar wandering data, was instrumental in verifying the theories of continental drift and plate tectonics in the 1960s and 70s. Some applications of paleomagnetic evidence to reconstructing histories of terranes have continued to arouse controversies. Paleomagnetic evidence also is used in constraining possible ages for rocks and processes and in reconstructions of the deformational histories of parts of the crust. Reversal magnetostratigraphy is often used to estimate the age of fossil and hominin bearing sites. Paleomagnetic studies are combined with geochronological methods to determine absolute ages for rocks in which the magnetic record is preserved. For igneous rocks such as basalt, commonly used methods include potassium argon and argon-argon geochronology.

Hideki Yukawa

Hideki Yukawa was a Japanese theoretical physicist and the first Japanese Nobel laureate. In 1949 he became a professor at Columbia University, the same year he received the Nobel Prize in Physics, after the discovery by Cecil Frank Powell, Giuseppe Occhialini and César Lattes of Yukawa's predicted pion in 1947.

Yukawa also worked on the theory of K-capture, in which a low energy electron is absorbed by the nucleus, after its initial prediction by G. C. Wick.

Born	23 January 1907, Tokyo, Japan
Died	8 September 1981 (age 74), Kyoto, Japan
Nationality	Japan
Fields	Theoretical Physics
Institutions	Osaka Imperial University, Kyoto Imperial University, Imperial University of Tokyo, Institute for Advanced Study, Columbia University
Alma mater	Kyoto Imperial University
Notable awards	Nobel Prize in Physics (1949)

PION (1949)

In particle physics, a pion is any of three subatomic particles: π^0, π^+ and π^-. Pions are the lightest mesons and play an important role in explaining low-energy properties of the strong nuclear force.

Pions are bosons with zero spin and are composed of first-generation quarks. In the quark model, an up quark and an anti-down quark compose a π^+, whereas a down quark and an anti-up quark compose the π^-, which are antiparticles of one another. The uncharged pions are combinations of an up quark with an anti-up quark or a down quark with an anti-down quark, have identical quantum numbers, and hence they are only found in superpositions. The lowest energy superposition of these is the π^0, which is its own antiparticle. Together, the pions form a triplet of isospin. Each pion has the isospin –1 ($I = 1$) and third component isospin is equal to its charge ($I_z = +1$, 0 or –1).

Electron capture is a decay mode for isotopes that will occur when there are too many protons in the nucleus of an atom and insufficient energy to emit a positron; however, it continues to be a viable decay mode for radioactive isotopes that can decay by positron emission. It is sometimes called inverse beta decay, though this term can also refer to the capture of a neutrino through a similar process. If the energy difference between the parent atom and the daughter atom is less than 1.022 Mev, positron emission is forbidden and electron capture is the sole decay mode. For example, Rubidium-83 will decay to Krypton-83 solely by electron capture (the energy difference is about 0.9 Mev).

In this case, one of the orbital electrons, usually from the K or L electron shell (K-electron capture, also K-capture, or L-electron capture, L-capture), is captured by a proton in the nucleus, forming a neutron and a neutrino.

$$p + e^- \rightarrow n + \gamma_e$$

Note that a free proton cannot normally be changed to a free neutron by this process. The proton and neutron must be part of a larger nucleus. Since the proton is changed to a neutron, the number of neutrons increases by 1, the number of protons decreases by 1, and the atomic mass number remains unchanged. By changing the number of protons, electron capture transforms the nuclide into a new element. The atom moves into an excited state with the inner shell missing an electron. When transiting to the ground state, the atom will emit an X-ray photon and/or Auger electrons.

Cecil Frank Powell

Cecil Frank Powell was a British physicist, and won Nobel Prize in Physics for his development of the photographic method of studying nuclear processes and for the resulting discovery of the pion (pi-meson), a heavy subatomic particle.

In 1949 Powell became a Fellow of the Royal Society and received the society's Hughes Medal the same year. In 1950 he was awarded the Nobel Prize for Physics for his development of the photographic method of studying nuclear processes and his discoveries regarding mesons made with this method. From 1952 Powell was appointed director of several expeditions to Sardinia and the Po Valley, Italy, utilizing high-altitude balloon flights.

Born	5 December 1903, Tonbridge, Kent, England
Died	9 August 1969 (age 65), Valsassina, Italy
Nationality	British
Fields	Physics
Institutions	University of Cambridge, University of Bristol
Alma mater	University of Cambridge
Doctoral advisor	C. T. R. Wilson, Ernest Rutherford
Known for	Photographic method, discovery of the pion
Notable awards	Nobel Prize in Physics (1950)

NUCLEAR EMULSION (1950)

In a Particle and Nuclear physics, a nuclear emulsion plate is a photographic plate with a particularly thick emulsion layer and with a very uniform grain size. Nuclear emulsions can be used to record and investigate fast charged particles like nucleons or mesons. After exposing and developing the plate, single particle tracks can be observed and measured using a microscope.

In 1937, Marietta Blau and Hertha Wambacher discovered nuclear *disintegration stars* due to spallation in nuclear emulsions that had been exposed to cosmic radiation at a height of 2300 m (7500 feet) above sea level.

Using nuclear emulsions exposed on high mountains, Cecil Frank Powell and coworkers discovered the pion in 1947.

PION

In particle physics, a pion (short for pi meson, denoted with π) is any of three subatomic particles: π^0, π^+, and π^-. Pions are the lightest mesons and they play an important role in explaining the low-energy properties of the strong nuclear force.

Pions are bosons with zero spin, and they are composed of first-generation quarks. In the quark model, an up quark and an anti-down quark make up a π^+, whereas a down quark and an anti-up quark make up the π^-, and these are the antiparticles of one another. The uncharged pions are combinations of an up quark with an anti-up quark or a down quark with an anti-down quark, have identical quantum numbers, and hence they are only found in superpositions. The lowest energy superposition of these is the π^0, which is its own antiparticle. Together, the pions form a triplet of isospin. Each pion has isospin (I = 1) and third component isospin equal to its charge (I_z = +1, 0 or −1).

The π^0 meson has a slightly smaller mass of 135.0 MeV/c² and a much shorter mean lifetime of 8.4×10^{-17}s. This pion decays in a electromagnetic force process. The main decay mode, with probability 0.98798, is into two photons (two gamma ray photons in this case): $\pi^0 \rightarrow 2\gamma$.

Its second most common decay mode, with probability 0.01198, is the so-called Dalitz decay into a photon and an electron-positron pair: $\pi^0 \rightarrow \gamma + e^- + e^+$. The rate at which pions decay is a prominent quantity in many sub-fields of particle physics, t such as chiral perturbation theory. This rate is parametrized by the pion decay constant ($f\pi$), which is about 90 Mev.

John Cockcroft

Sir John Douglas Cockcroft, was a British physicist. He received the Nobel Prize in Physics during 1951 for splitting the atomic nucleus, and was instrumental in the development of nuclear power.

Born	27 May 1897, Todmorden, England
Died	18 September 1967 (age 70), Cambridge, England
Nationality	United Kingdom
Fields	Physics
Institutions	Atomic Energy Research Establishment
Alma mater	Victoria University of Manchester, Manchester College of Technology, St. John's College, Cambridge
Academic advisors	Ernest Rutherford
Known for	Splitting the atom
Notable awards	Nobel Prize in Physics (1951)

COCKCROFT'S SUBATOMIC LEGACY: SPLITTING THE ATOM (1951)

The atom is a basic unit of matter that consists of a dense, central nucleus surrounded by a cloud of negatively charged electrons. The atomic nucleus contains a mix of positively charged protons and electrically neutral neutrons. The electrons of an atom are bound to the nucleus by the electromagnetic force. Likewise, a group of atoms can remain bound to each other, forming a molecule. An atom containing an equal number of protons and electrons is electrically neutral, otherwise it has a positive or negative charge and is an ion. An atom is classified according to the number of protons and neutrons in its nucleus: the number of protons determines the chemical element, and the number of neutrons determines the isotope of the element. In April 1932 John Cockcroft and Ernest Walton split the atom for the first time, at the Cavendish Laboratory in Cambridge in the UK. Only weeks earlier, James Chadwick, also in Cambridge, discovered the neutron. That same year, far away in California, Carl Anderson discovered the positron while working on cosmic rays. So 1932 was a veritable *annus mirabilis* in which experiments discovered, and worked with, nucleons; exploited Albert Einstein's relativity and energy-mass equivalence principle; took advantage of the newly emerging quantum mechanics and its prediction of tunnelling through potential barriers; and even verified the existence of antimatter predicted by Paul Dirac's relativistic quantum theory of the electron. It is hard to think of a more significant year in the annals of science. The experiment by Cockcroft and Walton split the nucleus at the heart of the atom with protons that were lower in energy than seemed possible, by virtue of quantum mechanical tunnelling, a phenomenon new to physics.

In 1928 George Gamow had applied the new quantum mechanics to show how particles could tunnel through potential barriers, and how this could explain the decay of nuclei through alpha emission. He also realized that tunnelling could lower the energy required for an incident positively charged particle to overcome the Coulomb barrier of a target nucleus. It was this insight that underpinned the commitment of Cockcroft and Walton. The entire sequence of events that led to the pioneering experiment (the specification of particle beam parameters based on contemporary theoretical application and phenomenology; the innovation and development of the necessary technologies to create such beams; and the use of the beams to do experiments on a subatomic scale to achieve a deeper understanding of the structure and function of matter) have been repeated many times as high-energy physics has advanced with the construction of accelerators to the current Standard Model of particles and forces. That Cockcroft realized the immense potential of accelerators in research, and in particular for progress in fundamental physics, is manifest in his instrumental role in later years to establish large accelerator laboratories, in particular CERN in 1954.

Ernest Walton

 Ernest Thomas Sinton Walton was an Irish physicist and Nobel laureate for his work with John Cockcroft with atom-smashing experiments done at Cambridge University in the early 1930s. Walton is the only Irishman to have won a Nobel Prize in science.

During the early 1930s Walton and John Cockcroft collaborated to build an apparatus that split the nuclei of lithium atoms by bombarding them with a stream of protons accelerated inside a high-voltage tube (700 kilovolts). The splitting of the lithium nuclei produced helium nuclei. This was experimental verification of theories about atomic structure that had been proposed earlier by Rutherford, George Gamow, and others. The successful apparatus is a type of particle accelerator now called the Cockcroft-Walton generator, helped to usher in an era of particle accelerator based experimental nuclear physics. It was this research at Cambridge that won Walton and Cockcroft the Nobel Prize in Physics in 1951.

Born	6 October 1903, Dungarvan, Ireland
Died	25 June 1995, Belfast, Northern Ireland
Nationality	Irish
Fields	Physics
Institutions	Trinity College Dublin, University of Cambridge, Methodist College Belfast
Doctoral advisor	Ernest Rutherford
Known for	The first disintegration of an atomic nucleus by artificially accelerated protons (splitting the atom)
Notable awards	Nobel Prize in Physics (1951)

THE FIRST DISINTEGRATION OF
AN ATOMIC NUCLEUS (1951)

Ernest Walton was an Irish experimental physicist who gained renown for achieving, with physicist John D. Cockcroft, the first artificial disintegration of an atomic nucleus, without the use of radioactive elements. Their breakthrough was accomplished by artificially accelerating a beam of protons and aiming it at a target of lithium, one of the lightest known metals. The resultant emission of alpha particles, that is, positively charged particles given off by certain radioactive substances, indicated not only that some protons had succeeded in penetrating the nuclei of the lithium atoms but also that they had somehow combined with the lithium atoms and had been transformed into something new. Although the process was not an efficient energy producer, the work of Walton and Cockcroft stimulated many theoretical and practical developments and influenced the whole course of nuclear physics. For their pioneering work, Walton and Cockcroft shared the 1951 Nobel Prize in Physics.

What was needed was a fundamentally different way of viewing the problem. Walton and his colleagues at the Cavendish were trying to accelerate electrons to a speed, sufficient to enable them to penetrate an atomic nucleus. Such high velocities were necessary, they believed, in order to counteract the repulsive charge of the nuclei. The speeding electrons, they figured, would literally bully their way through. However, achieving such high speeds was easier said than done. It required the application of enormous amounts of electricity, about four million volts, which at that time was impossible to generate in a discharge tube. A crucial breakthrough came in 1929, when the Russian physicist George Gamow visited the Cavendish laboratory. With physicist Niels Bohr in Copenhagen, he had worked out a wave-mechanical theory of the penetration of particles, in which they believed particles tunneled through rather than over potential barriers. This meant that particles propelled by about 500,000 volts, as opposed to millions, could possibly permeate the barrier and enter the nucleus if present in sufficiently large numbers.

Felix Bloch

Felix Bloch was a Swiss physicist, working mainly in the U.S. In 1946 he proposed the Bloch equations which determine the time evolution of nuclear magnetization. He and Edward Mills Purcell were awarded the 1952 Nobel Prize for their development of new ways and methods for nuclear magnetic precision measurements.

Born	23 October 1905, Zürich, Switzerland
Died	10 September 1983 (age 77), Zürich, Switzerland
Nationality	Swiss
Fields	Physics
Institutions	Stanford University
Alma mater	ETH Zürich and University of Leipzig
Doctoral advisor	Werner Heisenberg
Known for	NMR, Bloch wall, Bloch's Theorem
Notable awards	Nobel Prize in Physics (1952)

NUCLEAR MAGNETIC RESONANCE (1952)

Nuclear Magnetic Resonance (NMR) is a property that magnetic nuclei have in a magnetic field and applied electromagnetic (EM) pulse or pulses, which cause the nuclei to absorb energy from the EM pulse and radiate this energy back out. The energy radiated back out is at a specific resonance frequency which depends on the strength of the magnetic field and other factors. This allows the observation of specific quantum mechanical magnetic properties of an atomic nucleus. Many scientific techniques exploit NMR phenomena to study molecular physics, crystals and non-crystalline materials through NMR spectroscopy. NMR is also routinely used in advanced medical imaging techniques, such as in magnetic resonance imaging (MRI). Nuclear magnetic resonance was first described and measured in molecular beams by Isidor

Rabi in 1938, and in 1944, Rabi was awarded the Nobel Prize in physics for this work. In 1946, Felix Bloch and Edward Mills Purcell expanded the technique for use on liquids and solids, for which they shared the Nobel Prize in physics in 1952. Purcell had worked on the development and radar applications during World War II at Massachusetts Institute of Technology's Radiation Laboratory. His work during that project on the production and detection of RF energy, and on the absorption of such RF energy by matter, preceded his discovery of NMR. They noticed that magnetic nuclei, like ^1H and ^{31}P, could absorb RF energy when placed in a magnetic field of a strength specific to the identity of the nuclei. When this absorption occurs, the nucleus is described as being *in resonance*. Different atomic nuclei within a molecule resonate at different frequencies for the same magnetic field strength. The observation of such magnetic resonance frequencies of the nuclei present in a molecule allows any trained user to discover essential, chemical and structural information about the molecule. The development of nuclear magnetic resonance as a technique of analytical chemistry and biochemistry parallels the development of electromagnetic technology and its introduction into civilian use.

BLOCH WALL

A Bloch wall is a narrow transition region at the boundary between magnetic domains, over which the magnetization changes from its value in one domain to that in the next. The magnetization rotates through the plane of the wall unlike the Néel wall where the magnetization rotates in the plane of the wall. For example, a Bloch wall is formed at the center line of a bar magnet where the domains switch north/south direction. Bloch walls are named after the physicist Felix Bloch.

Edward Mills Purcell

Edward Mills Purcell was an American physicist who shared the 1952 Nobel Prize in Physics for his independent discovery (published 1946) of nuclear magnetic resonance in liquids and in solids. Nuclear magnetic resonance (NMR) has become widely used to study the molecular structure of pure materials and the composition of mixtures.

Born	30 August 1912, Taylorville, Illinois, USA
Died	7 March 1997 (age 84), Cambridge, Massachusetts, USA
Nationality	United States
Fields	Physics
Institutions	Harvard University, MIT
Alma mater	Purdue University, Harvard University
Doctoral advisor	Kenneth Bainbridge
Academic advisors	John Van Vleck
Doctoral students	Nicolaas Bloembergen, George Pake, George Benedek
Known for	Nuclear magnetic resonance (NMR), Smith-Purcell effect 21 cm line
Notable awards	Nobel Prize in Physics (1952)

NUCLEAR MAGNETIC RESONANCE IN SOLIDS AND LIQUIDS (1952)

All stable isotopes that contain an odd number of protons and/or of neutrons have an intrinsic magnetic moment and angular momentum, in other words a nonzero spin, while all nuclides with even numbers of both have spin 0. The most commonly studied nuclei are ^1H (the most NMR-sensitive isotope after the radioactive ^3H) and ^{13}C, although nuclei from isotopes of many other elements (e.g. ^2H, ^{10}B, ^{11}B, ^{14}N, ^{15}N, ^{17}O, ^{19}F, ^{23}Na, ^{29}Si, ^{31}P, ^{35}Cl, ^{113}Cd, ^{129}Xe, ^{195}Pt) are studied by high-field NMR spectroscopy as well. A key feature of NMR is that the resonance frequency of a particular substance is directly proportional to the strength of the applied magnetic field. It is this feature that is exploited in imaging techniques; if a sample is placed in a non-uniform magnetic field then the resonance frequencies of the sample's nuclei depend on where in the field they are located. Since the resolution of the imaging techniques depends on how big the gradient of the field is, many efforts are made to develop more powerful magnets, often using superconductors. The effectiveness of NMR can also be improved using hyper polarization, and/or using two-dimensional, three-dimensional and higher dimension multi-frequency techniques.

The principle of NMR usually involves two sequential steps:

- The alignment (polarization) of the magnetic nuclear spins in an applied, constant magnetic field H_0.
- The perturbation of this alignment of the nuclear spins by employing an electromagnetic, usually radio-frequency (RF) pulse. The required perturbing frequency is dependent upon the static magnetic field (H_0) and the nuclei of observation.

The two fields are usually chosen to be perpendicular to each other as this maximizes the NMR signal strength. The resulting response by the total magnetization (M) of the nuclear spins is the phenomenon that is exploited in NMR spectroscopy and magnetic resonance imaging. Both use intense applied magnetic fields (H_0) in order to achieve dispersion and very high stability to deliver spectral resolution, the details of which are described by chemical shifts, the Zeeman effect, and Knight shifts in metals.

Frits Zernike

 Frits Zernike was a Dutch physicist and winner of the Nobel prize for physics in 1953 for his invention of the phase contrast microscope, an instrument that permits the study of internal cell structure without the need to stain and thus kill the cells.

Born	16 July 1888, Amsterdam, Netherlands
Died	10 March 1966 (age 77), Amersfoort, Netherlands
Nationality	Netherlands
Fields	Physics
Institutions	Groningen University
Alma mater	University of Amsterdam
Doctoral students	Christoffel Bouwkamp, Herman de Boer, Bernard Nijboer
Known for	Ornstein-Zernike equation, Zernike polynomials phase contrast microscopy
Notable awards	Rumford Medal (1952), Nobel Prize in Physics (1953)

PHASE CONTRAST MICROSCOPY (1953)

Phase contrast microscopy is an optical microscopy illumination technique in which small phase shifts in the light passing through, a transparent specimen are converted into amplitude or contrast changes in the image.

A phase contrast microscope does not require staining to view the slide. This type of microscope made it possible to study the cell cycle.

As light travels through a medium other than vacuum, interaction with this medium causes its amplitude and phase to change in a way which depends on properties of the medium. Changes in amplitude give rise to familiar absorption of light which gives rise to colours which is wavelength dependent. The human eye measures only the energy of light arriving on the retina, so changes in phase are not easily observed, yet often these changes in phase carry a large amount of information.

The same holds in a typical microscope, i.e. although the phase variations introduced by the sample are preserved by the instrument this information is lost in the process which measures the light. In order to make phase variations observable, it is necessary to combine the light passing through the sample with a reference so that the resulting interference reveals the phase structure of the sample.

This was first realized by Frits Zernike during his study of diffraction gratings. During these studies he appreciated both that it is necessary to interfere with a reference beam, and that to maximise the contrast achieved with the technique, it is necessary to introduce a phase shift to this reference so that the no-phase-change condition gives rise to completely destructive interference.

He later realised that the same technique can be applied to optical microscopy. The necessary phase shift is introduced by rings etched accurately onto glass plates so that they introduce the required phase shift when inserted into the optical path of the microscope. When in use, this technique allows phase of the light passing through the object under study to be inferred from the intensity of the image produced by the microscope. This is the phase-contrast technique.

In optical microscopy many objects such as cell parts in protozoans, bacteria and sperm tails are essentially fully transparent unless stained. Staining is a difficult and time consuming procedure which sometimes, but not always, destroys or alters the specimen. The difference in densities and composition within the imaged objects however often give rise to changes in the phase of light passing through them, hence they are sometimes called phase objects. Using the phase-contrast technique makes these structures visible and allows their study with the specimen still alive.

Max Born

Max Born was a German born physicist and mathematician who was instrumental in the development of quantum mechanics. He also made contributions to solid-state physics and optics and supervised the work of a number of notable physicists in the 1920s and 30s. Born won the 1954 Nobel Prize in Physics.

Born	11 December 1882, Breslau, Germany
Died	5 January 1970 (age 87), Göttingen, Germany
Residence	Göttingen, Germany
Citizenship	Germany/United Kingdom
Nationality	German
Fields	Physicist
Institutions	University of Frankfurt am Main, University of Göttingen, University of Edinburgh
Alma mater	University of Göttingen
Doctoral students	Victor Frederick Weisskopf, J. Robert Oppenheimer, Lothar Wolfgang Nordheim, Max Delbrück, Walter Elsasser, Friedrich Hund, Pascual Jordan, Maria Goeppert-Mayer, Herbert S. Green, Cheng Kaijia, Siegfried Flügge, Edgar Krahn, Maurice Pryce, Antonio Rodríguez, Bertha Swirles, Paul Weiss, Peng Huanwu
Academic advisors	Carl Runge, Joseph Larmor, J. J. Thomson
Known for	Born-Haber cycle, Born rigidity, Born approximation, Born-Infeld theory, Born-Oppenheimer approximation, Born's Rule, Born-Landé equation
Notable awards	Nobel Prize in Physics (1954)

QUANTUM MECHANICS/WAVE FUNCTION (1954)

Quantum mechanics (QM) or Quantum Physics, is a branch of physics describing much of the behavior of energy and matter at the atomic and subatomic scales. The name derives from the observation that some physical quantities, such as the angular momentum or, more generally, the action of, for example, an electron bound into an atom or molecule—can be changed only by discrete amounts, or quanta as multiples of the Planck constant, rather than being capable of varying by any amount. An electron bound in an atomic orbital has quantized values of angular momentum while an unbound electron does not exhibit quantized energy levels, but is associated with a quantum mechanical wavelength.

In the context of QM, the wave-particle duality of energy and matter at the atomic scale provides a unified view of the behavior of particles such as photons and electrons and other atomic-scale particles. Historically, the earliest versions of QM were formulated in the first decade of the 20th century at around the same time as the atomic theory and the corpuscular theory of light as updated by Einstein first came to be widely accepted as scientific fact. QM underwent a significant re-formulation in the mid 1920s away from old quantum theory with the acceptance of the Copenhagen interpretation.

The mathematical formulations of quantum mechanics are abstract and the implications are often non-intuitive. The center-piece of the mathematical system of the Schrödinger picture is the wave function. The wave function is a mathematical function of time and space that can provide information about the position and momentum of a particle, but only as probabilities, as dictated by the constraints imposed by the uncertainty principle related with the Heisenberg picture. Mathematical manipulations of the wave function usually involve the bracket notation, which requires an understanding of complex numbers and linear functionals. Many of the results of QM can only be expressed mathematically and do not have models that are as easy to visualize as those of classical mechanics.

Walther Bothe

 Walther Wilhelm Georg Bothe was a German nuclear physicist. He developed and applied coincidence methods to the study of nuclear reactions, the Compton Effect, cosmic rays, and the wave-particle duality of radiation, for which he received the Nobel Prize in Physics in 1954.

Born	8 January 1891, Oranienburg, Germany
Died	8 February 1957, Heidelberg, Germany
Nationality	Germany
Fields	Physics, Mathematics, Chemistry
Institutions	University of Berlin, University of Giessen, University of Heidelberg Max Planck Institute for Medical Research
Alma mater	University of Berlin
Doctoral students	Hans Ritter von Baeyer
Academic advisors	Max Planck
Known for	Coincidence circuit
Notable awards	Nobel Prize in Physics (1954), Max Planck Medal (1953)

Walther Wilhelm Georg Bothe shared the Nobel Prize in Physics in 1954 with Max Born.

In 1913, he joined the newly created Laboratory for Radioactivity at the Reich Physical and Technical Institute (PTR), where he remained until 1930, the later few years as the director of the laboratory. He served in the military during World War I from 1914, and he was a prisoner of war of the Russians, returning to Germany in 1920. Upon his return to the laboratory, he developed and applied coincidence methods to the study of nuclear reactions, the Compton effect, Cosmic rays, and the wave-particle duality of radiation, for which he received the Nobel Prize.

In 1930 he became professor and director of the physics department at the University of Giessen. In 1932, he became director of the Physical and Radiological Institute at the

University of Heidelberg. He was driven out of this position by elements of the deutsche Physik movement. To preclude his emigration from Germany, he was appointed director of the Physics Institute of the Kaiser Wilhelm Institute for Medical Research (KWImF) in Heidelberg. There, he built the first operational cyclotron in Germany. Furthermore, he became a principal in the German nuclear energy project, also known as the Uranium Club, which was started in 1939 under the supervision of the Army Ordnance Office.

In 1946, in addition to his directorship of the Physics Institute at the KWImf, he was reinstated as a professor at the University of Heidelberg. From 1956 to 1957, he was a member of the Nuclear Physics Working Group in Germany.

In the year after Bothe's death, his Physics Institute at the KWImF was elevated to the status of a new institute under the Max Planck Society and it then became the Max Planck Institute for Nuclear Physics. Its main building was later named Bothe laboratory.

COINCIDENCE CIRCUIT (1954)

In physics, a coincidence circuit is an electronic device with one output and two (or more) inputs. The output is activated only when signals are received within a time window accepted as *at the same time* and in parallel at both inputs. Coincidence circuits are widely used in particle physics experiments and in other areas of science and technology.

Walther Bothe shared the Nobel Prize for Physics in 1954 for his discovery of the method of coincidence and the discoveries subsequently made by it. Bruno Rossi invented the electronic coincidence circuit for implementing the coincidence method.

Willis Lamb

Willis Eugene Lamb, Jr. was an American physicist who won the Nobel Prize in Physics in 1955 for his discoveries concerning the fine structure of the hydrogen spectrum. Lamb and Polykarp Kusch were able to precisely determine certain electromagnetic properties of the electron. Lamb was a professor at the University of Arizona College of Optical Sciences.

Born	12 July 1913, Los Angeles, California, U.S.
Died	15 May 2008 (age 94), Tucson, Arizona, U.S.
Nationality	United States
Fields	Physics
Institutions	University of Arizona, University of Oxford Yale, Columbia, Stanford
Alma mater	University of California, Berkeley
Doctoral students	Theodore Maiman, Marlan Scully, Balázs László Gyorffy, Frederick Hopf, Murray Sargent III, Stanley L. Kaufman, David Mader, Ralph Jacobs
Academic advisors	J. Robert Oppenheimer
Known for	Lamb shift, Laser Theory, Quantum Optics
Notable awards	Nobel Prize in Physics (1955)

LAMB SHIFT (1955)

In physics, the Lamb shift, named after Willis Lamb, is a small difference in energy between two energy levels $^2S_{1/2}$ and $^2P_{1/2}$ of the hydrogen atom in quantum electrodynamics (QED). According to Dirac, the $^2S_{1/2}$ and $^2P_{1/2}$ orbitals should have the same energies. However, the interaction between the electron and the vacuum causes a tiny energy shift on $^2S_{1/2}$. Lamb and Robert Retherford measured this shift in 1947, and this measurement provided the stimulus for renormalization theory to handle the divergences. It was the harbinger of modern QED as developed by Schwinger, Feynman, and Dyson.

In 1947 Willis Lamb and Robert Retherford carried out an experiment using microwave techniques to stimulate radio-frequency transitions between $^2S_{1/2}$ and $^2P_{1/2}$ levels of hydrogen. By using lower frequencies than for optical transitions the Doppler broadening could be neglected (Doppler broadening is proportional to the frequency). The energy difference Lamb and Retherford found was a rise of about 1000 MHz of the $^2S_{1/2}$ level above the $^2P_{1/2}$ level.

This particular difference is a one-loop effect of quantum electrodynamics, and can be interpreted as the influence of virtual photons that have been emitted and re-absorbed by the atom. In quantum electrodynamics the electromagnetic field is quantized and, like the harmonic oscillator in quantum mechanics, its lowest state is not zero. Thus, there exist small zero-point oscillations that cause the electron to execute rapid oscillatory motions. The electron is smeared out and the radius is changed from r to $r + \delta r$.

In 1947, Hans Bethe was the first to explain the Lamb shift in the hydrogen spectrum, and he thus laid the foundation for the modern development of quantum electrodynamics. The Lamb shift currently provides a measurement of the fine-structure constant α to better than one part in a million, allowing a precision test of quantum electrodynamics. A different perspective relates Zitterbewegung to the Lamb shift.

Zitterbewegung (trembling motion) is a theoretical rapid motion of elementary particles, in particular electrons, that obey the Dirac equation. The existence of such motion was first proposed by Erwin Schrödinger in 1930 as a result of his analysis of the wave packet solutions of the Dirac equation for relativistic electrons in free space, in which an interference between positive and negative energy states produces what appears to be a fluctuation (at the speed of light) of the position of an electron around the median, with a circular frequency of,

$\dfrac{2mc^2}{h}$ or approximately 1.6×10^{21} Hz.

Polykarp Kusch

Polykarp Kusch was a German-American physicist. In 1955 he was jointly awarded the Nobel Prize for Physics with Willis Eugene Lamb for his accurate determination that the magnetic moment of the electron was greater than its theoretical value, thus leading to reconsideration of magnetic moment and innovations in quantum electrodynamics.

Born	26 January 1911, Blankenburg, Germany
Died	20 March 1993 (aged 82)
Institutions	University of Texas at Dallas, Columbia University
Alma mater	University of Illinois, Case Western Reserve University
Doctoral advisor	F. Wheeler Loomis
Notable students	Gordon Gould
Known for	Measured the magnetic moment of the electron
Notable awards	Nobel Prize in Physics (1955)

ELECTRON MAGNETIC DIPOLE MOMENT (1955)

In atomic physics, the electron magnetic dipole moment is the magnetic moment of an electron caused by its intrinsic property of spin.

The electron is a charged particle. Its angular momentum comes from two types of rotation: spin and orbital motion. From classical electrodynamics, a rotating electrically charged body creates a magnetic dipole with magnetic poles of equal magnitude but opposite polarity. This analogy holds as an electron indeed behaves like a tiny bar magnet. One consequence is that an external magnetic field exerts a torque on the electron magnetic moment depending on its orientation with respect to the field.

If the electron is visualized as a classical charged particle literally rotating about an axis with angular momentum $\vec{L}$, its magnetic dipole moment $\vec{\mu}$ is given by:

$$\vec{\mu} = \frac{-e}{2m_e}\vec{L}$$

here the charge is $-e$, where e is the elementary charge. The mass is the electron rest mass m_e. Note that the angular momentum $\overset{\circledR}{L}$ in this equation may be the spin angular momentum, the orbital angular momentum, or the total angular momentum. It turns out the classical result is off by a proportional factor for the spin magnetic moment. As a result, the classical result is corrected by multiplying it with a correction factor:

$$\vec{\mu} = g\frac{-e}{2m_e}\vec{L}$$

the dimensionless correction factor g is known as the g-factor. Finally, it is customary to express the magnetic moment in terms of the Planck constant and the Bohr magneton:

$$\vec{\mu} = -g\mu B\frac{\vec{L}}{\hbar}$$

here μ_B is the Bohr magneton and $\hbar$ is the reduced Planck constant.

For the electron spin, the most accurate value for the spin g-factor has been experimentally determined to have the value $2.00231930419922 \pm (1.5 \times 10^{-12})$. Note that it is only two thousandths larger than the value from Dirac equation. The small correction is known as the anomalous magnetic dipole moment of the electron; it arises from the electron's interaction with virtual photons in quantum electrodynamics. In fact, one famous triumph of the Quantum Electrodynamics theory is the accurate prediction of the electron g-factor. The most accurate value for the electron magnetic moment is $-928.476377 \times 10^{-26} \pm 0.000023 \times 10^{-26}$ J/T.

William Shockley

William Bradford Shockley was an American physicist and inventor. Along with John Bardeen and Walter Houser Brattain, Shockley co-invented the transistor, for which all three were awarded the 1956 Nobel Prize in Physics.

Born	13 February 1910, London, England, United Kingdom
Died	12 August 1989 (age 79), Stanford, California, United States
Nationality	American
Institutions	Bell Labs, Shockley Semiconductor, Stanford
Alma mater	Caltech, MIT
Doctoral advisor	John C. Slater
Known for	Co-inventor of the transistor
Notable awards	Nobel Prize in Physics (1956)

SOLID-STATE TRANSISTOR (1956)

Shortly after the end of the war in 1945, Bell Labs formed a Solid State Physics Group, led by Shockley and chemist Stanley Morgan; other personnel including John Bardeen and Walter Brattain, physicist Gerald Pearson, chemist Robert Gibney, electronics expert Hilbert Moore and several technicians. Their assignment was to seek a solid-state alternative to fragile glass vacuum tube amplifiers. Their first attempts were based on Shockley's ideas about using an external electrical field on a semiconductor to affect its conductivity. These experiments failed every time in all sorts of configurations and materials. The group was at a standstill until Bardeen suggested a theory that invoked surface states that prevented the field from penetrating the semiconductor. The group changed its focus to study these surface states and they met almost daily to discuss the work. The rapport of the group was excellent, and ideas were freely exchanged.

By the winter of 1946 they had enough results that Bardeen submitted a paper on the surface states to *Physical Review*. Brattain started experiments to study the surface states through observations made while shining a bright light on the semiconductor's surface. This led to several more papers (one of them co-authored with Shockley), which estimated the density of the surface states to be more than enough to account for their failed experiments. The pace of the work picked up significantly when they started to surround point contacts between the semiconductor and the conducting wires with electrolytes. Moore built a circuit that allowed them to vary the frequency of the input signal easily. Finally they began to get some evidence of power amplification when Pearson, acting on a suggestion by Shockley, put a voltage on a droplet of glycol borate (a viscous chemical that did not evaporate) placed across a P-N junction.

December 1947 was Bell Labs' Miracle Month, when Bardeen and Brattain working without Shockley–succeeded in creating a point-contact transistor that achieved amplification. By the next month, Bell Lab's patent attorneys started to work on the patent applications.

Bell Labs attorneys soon discovered that Shockley's field effect principle had been anticipated and patented in 1930 by Julius Lilienfeld, who filed his MESFET-like patent in Canada on 22 October 1925. Although the patent appeared breakable (it could not work) the patent attorneys based one of its four patent applications only on the Bardeen-Brattain point contact design. Three others covered the electrolyte-based transistors with Bardeen, Gibney and Brattain as the inventors. Shockley's name was not on any of these patent applications. This angered Shockley, who thought his name should also be on the patents because the work was based on his field effect idea. He even made efforts to have the patent written only in his name, and told Bardeen and Brattain of his intentions.

At the same time he secretly continued his own work to build a different sort of transistor based on junctions instead of point contacts; he expected this kind of design would be more likely to be commercially viable.

John Bardeen

John Bardeen was an American physicist and electrical engineer, the first person to have won the Nobel Prize in Physics twice: first in 1956 with William Shockley and Walter Brattain for the invention of the transistor; and again in 1972 with Leon Neil Cooper and John Robert Schrieffer for a fundamental theory of conventional superconductivity known as the BCS theory.

Born	23 May 1908, Madison, Wisconsin, USA
Died	30 January 1991 (age 82), Boston, Massachusetts
Nationality	American
Fields	Physics
Institutions	Bell Labs, University of Minnesota, University of Illinois at Urbana-Champaign
Alma mater	University of Wisconsin-Madison, Princeton University
Doctoral advisor	Eugene Wigner
Doctoral students	Nick Holonyak, John Schrieffer
Known for	Transistor, BCS theory
Notable awards	Nobel Prize in Physics (1956), IEEE Medal of Honor (1971), Nobel Prize in Physics (1972)

THE INVENTION OF THE TRANSISTOR (1956)

In the spring of 1947, William Shockley set Brattain and Bardeen to a task to explain why an amplifier he had devised didn't work. At the heart of the amplifier was a crystal of silicon. They would switch to germanium after some months. To figure out what was going on, Bardeen had to remember some of the quantum mechanics research that he had done on semiconductors while he was completing his PhD at Princeton University. Bardeen had also come up with some new theories himself. By observing Brattain's experiments, Bardeen realized that everyone had been falsely assuming electrical current traveled through all parts of the germanium in a similar way. The electrons behaved differently at the surface of the metal. If they could control what was happening at the surface, the amplifier should work.

On 23 December 1947, Bardeen and Brattain working without Shockley succeeded in creating a point-contact transistor that achieved amplification. By the next month, Bell Labs' patent attorneys started to work on the patent applications.

At the same time, Shockley secretly continued his own work to build a different sort of transistor based on junctions instead of point contacts; he expected this kind of design would be more likely to be viable commercially. Shockley worked furiously on his magnum opus, *Electrons and Holes in Semiconductors*, which was finally published as a 558-page treatise in 1950. In it, Shockley worked out the critical ideas of drift and diffusion and the differential equations that govern the flow of electrons in solid state crystals. Shockley's diode equation is also described.

Shockley was dissatisfied with certain parts of the explanation for how the point contact transistor worked and conceived of the possibility of minority carrier injection. This led Shockley to ideas for what he called a sandwich transistor. This resulted in the junction transistor, which was announced at a press conference on 4 July, 1951. Shockley obtained a patent for this invention on 25 September, 1951. Different fabrication methods for this device were developed but the diffused-base method became the method of choice for many applications. It soon eclipsed the point contact transistor, and it and its offspring became overwhelmingly dominant in the marketplace for many years. Shockley continued as a group head to lead much of the effort at Bell Labs to improve it and its fabrication for two more years.

Walter Houser Brattain

Walter Houser Brattain was an American physicist at Bell Labs who, along with John Bardeen and William Shockley, invented the transistor. They shared the 1956 Nobel Prize in Physics for their invention. He devoted much of his life to research on surface states.

Born	10 February 1902, Amoy, China
Died	13 October 1987, Seattle, Washington, USA
Nationality	United States
Fields	Physicist,
Institutions	Whitman College, Bell Laboratories
Alma mater	Whitman College, University of Oregon, University of Minnesota
Doctoral advisor	John Torrence Tate, Sr.
Known for	Transistor
Notable awards	Stuart Ballantine Medal (1952), Nobel Prize in Physics (1956)

TRANSISTOR (1956)

Brattain's concerns at Bell Laboratories in the years before World War II were first in the surface physics of tungsten and later in the surfaces of the semiconductors cuprous oxide and silicon. During World War II Brattain devoted his time to developing methods of submarine detection under a contract with the National Defense Research Council at Columbia University.

Following the war, Brattain returned to Bell Laboratories and soon joined the semiconductor division of the newly organized Solid State Department of the laboratories. William Shockley was the director of the semiconductor division, and early in 1946 he initiated a general investigation of semiconductors that was intended to produce a practical solid state amplifier.

Crystals of pure semiconductors are very poor conductors at ambient temperatures because the energy that an electron must have in order to occupy a conduction energy level is considerably greater than the thermal energy available to an electron in such a crystal. Heating a semiconductor can excite electrons into conduction states, but it is more practical to increase conductivity by adding impurities to the crystal. A crystal may be doped with a small amount of an element having more electrons than the semiconductor, and those excess electrons will be free to move through the crystal; such a crystal is an n-type semiconductor. One may also add to the crystal a small amount of an element having fewer electrons than the semiconductor, and the electron vacancies, or holes, so introduced will be free to move through the crystal like positively-charged electrons; such a doped crystal is a p-type semiconductor.

At the surface of a semiconductor the level of the conduction band can be altered, which will increase or decrease the conductivity of the crystal. Junctions between metals and n-type or p-type semiconductors, or between the two types of semiconductors, have asymmetric conduction properties, and semiconductor junctions can therefore be used to rectify electrical currents. In a rectifier, a voltage bias that produces a current flow in the low-resistance direction is a forward bias, while a bias in the opposite direction is a reverse bias.

Semiconductor rectifiers were familiar devices by the end of World War II, and Shockley hoped to produce a new device that would have a variable resistance and hence could be used as an amplifier. He proposed a design in which an electric field was applied across the thickness of a thin slab of a semiconductor. The conductivity of the semiconductor changed only by a small fraction of the expected amount when the field was applied, which John Bardeen suggested was due to the existence of energy states for electrons on the surface of the semiconductor.

Chen Ning Yang

Chen-Ning Franklin Yang born on 1 October 1922, is a Chinese-American physicist who worked on statistical mechanics and particle physics. He, together with Tsung-dao Lee, received the 1957 Nobel Prize in physics for their work on parity nonconservation of weak interaction. Yang naturalized as United States citizen in 1964.

Born	1 October 1922 (age 87), Hefei, Anhui, China.
Residence	China
Nationality	United States(1964)
Fields	Physics
Institutions	Institute for Advanced Study, State University of New York at Stony Brook, Chinese University of Hong Kong, Tsinghua University University of Chicago
Alma mater	National Southwestern Associated University, Tsinghua University, University of Chicago
Doctoral advisor	Edward Teller
Doctoral students	Bill Sutherland
Known for	Parity violation, Yang-Mills theory, Yang-Baxter equation
Notable awards	Nobel Prize in Physics (1957), Rumford Prize (1980), National Medal of Science (1986), Benjamin Franklin Medal (1993), Albert Einstein Medal (1995)

PARITY (1957)

In physics, a parity transformation is the flip in the sign of *one* spatial coordinate. In three dimensions, it is also commonly described by the simultaneous flip in the sign of all three spatial coordinates:

$$P : \begin{pmatrix} x \\ y \\ z \end{pmatrix} \rightarrow \begin{pmatrix} -x \\ -y \\ -z \end{pmatrix}$$

A 3 × 3 matrix representation of P would have determinant equal to –1, and hence cannot reduce to a rotation which has a determinant equal to 1. The corresponding mathematical notion is that of a point reflection.

In a two-dimensional plane, parity is not a simultaneous flip of all coordinates, which would be the same as a rotation by 180°. It is important that the determinant of the P matrix be -1, which does not happen for 180° rotation in 2D where a parity transformation flips the sign of *either* x or y, not *both*.

Parity violation: Although parity is conserved in electro-magnetism, strong interactions and gravity, it turns out to be violated in weak interactions. The Standard Model incorporates parity violation by expressing the weak interaction as a chiral gauge interaction. Only the left-handed components of particles and right-handed components of antiparticles participate in weak interactions in the Standard Model. This implies that parity is not a symmetry of our universe, unless a hidden mirror sector exists in which parity is violated in the opposite way.

The history of the discovery of parity violation is interesting. It was suggested several times and in different contexts that parity might not be conserved, but in the absence of compelling evidence these suggestions were not taken seriously. A careful review by theoretical physicists Tsung-Dao Lee and Chen Ning Yang went further, showing that while parity conservation had been verified in decays by the strong or electromagnetic interactions, it was untested in the weak interaction. They proposed several possible direct experimental tests. They were almost ignored, but Lee was able to convince his Columbia colleague Chien-Shiung Wu to try it. She needed special cryogenic facilities and expertise, so the experiment was done at the National Bureau of Standards.

Tsung-Dao Lee

Tsung-Dao Lee (T.D. Lee), is a Chinese-born American physicist, well known for his work on parity violation, the Lee Model, particle physics, relativistic heavy ion (RHIC) physics, nontopological solitons and soliton stars.

In 1957, Lee, at age 30 or 31, depending on announcement date or ceremony date, with C. N. Yang won the Nobel Prize in Physics for their work on the violation of parity law in weak interaction, which Chien-Shiung Wu experimentally verified. Lee is the second youngest Nobel laureate, after W. L. Bragg who won the prize at the age of 25, with his father W. H. Bragg in 1915. Lee and Yang were the first Chinese Laureates. Since naturalized as American citizen in 1962, Lee thus is also the youngest American who has ever won a Nobel Prize. In December 2007, Lee was, again, invited to the Nobel Prize ceremony in Stockholm by the Royal Swedish Academy of Sciences, half a century after winning his Nobel Prize.

Born	24 November 1926 (age 83), Shanghai, China
Citizenship	Republic of China, United States (1962)
Fields	Physics
Institutions	Columbia University
Alma mater	Zhejiang University, National Southwestern Associated University, University of Chicago
Doctoral advisor	Enrico Fermi
Known for	Parity violation, Lee model, non-topological solitons, particle physics, relativistic heavy ion (RHIC) Physics
Notable awards	Nobel Prize in Physics (1957), Albert Einstein Award (1957)

POSSIBLE EIGEN VALUES (1957)

In quantum mechanics, space time transformations act on quantum states. The parity transformation, P, is a unitary operator in quantum mechanics, acting on a state ψ as follows: $P \psi (r) = \psi (-r)$. One must have $P^2 \psi (r) = e^{i \psi} \psi (r)$, since an overall phase is unobservable.

The operator P^2, which reverses the parity of a state twice, leaves the space time invariant and so is an internal symmetry which rotates its eigen states by phases $e^{i \psi}$. If P^2 is an element $e^{i Q}$ of a continuous U(1) symmetry group of phase rotations then $e^{-i Q/2}$ is part of this U(1) and so is also a symmetry. In particular we can define $P' = P e^{-i Q/2}$ which is also a symmetry and so we can choose to call P' our parity operator instead of P. Notice that $P'^2 = 1$ and so P' has eigen values ± 1. However, when no such symmetry group exists, it may be that all parity transformations have some eigen values which are phases other than ± 1.

When parity generates the Abelian group Z_2, one can always take linear combinations of quantum states such that they are either even or odd under parity. Thus the parity of such states is ± 1. The parity of a multiparticle state is the product of the parities of each state; in other words parity is a multiplicative quantum number.

In quantum mechanics, Hamiltonians are invariant under a parity transformation if P commutes with the Hamiltonian. In non-relativistic quantum mechanics, this happens for any potential which is scalar, i.e. $V = V(r)$, hence the potential is spherically symmetric. The following facts can be easily proven:

- If $|A>$ and $|B>$ have the same parity, then $<A| X |B> = 0$ where X is the position operator.
- For a state $|L, m>$ of orbital angular momentum L with z-axis projection m, $P |L, m> = (-1)^L |L, m>$.
- If $[H, P] = 0$, then transitions only occur between states of opposite parity.
- If $[H, P] = 0$, then a non-degenerate eigen state of H is also an eigen state of the parity operator, i.e. a non-degenerate eigenfunction of H is either invariant to P or is changed in sign by P.

Pavel Cherenkov

Pavel Alekseyevich Cherenkov was a Soviet physicist who won the Nobel Prize in physics in 1958 for the discovery of the Cherenkov radiation he made in 1934.

Born	15 July 1904, Voronezh Oblast, Russian Empire
Died	6 January 1990 (age 85), Moscow, Russia
Nationality	Russian
Fields	Nuclear physics
Institutions	Lebedev Physical Institute
Alma mater	Voronezh State University
Doctoral advisor	Sergey Vavilov
Known for	Cherenkov radiation
Notable awards	Nobel Prize in Physics (1958)

CHERENKOV RADIATION (1958)

Cherenkov radiation (also spelled Cerenkov or Cerenkov) is electromagnetic radiation emitted when a charged particle (such as an electron) passes through an insulator at a constant speed greater than the speed of light in that medium. The charged particles polarize the molecules of that medium, which then turn back rapidly to their ground state, emitting prompt radiation. The characteristic blue glow of nuclear reactors is due to Cherenkov radiation. It is named after Russian scientist Pavel Alekseyevich Cherenkov, the 1958 Nobel Prize winner who was the first to characterize it rigorously.

The frequency spectrum of Cherenkov radiation by a particle is given by the Frank-Tamm formula. Unlike fluorescence or emission spectra that have characteristic spectral peaks, Cherenkov radiation is continuous. Around the visible spectrum, the relative intensity per unit frequency is approximately proportional to the frequency. That is, higher frequencies (shorter wavelengths) are more intense in Cherenkov radiation. This is why visible Cherenkov radiation is observed to be brilliant blue. In fact, most Cherenkov radiation is in the ultraviolet spectrum—it is only with sufficiently accelerated charges that it even becomes visible; the sensitivity of the human eye peaks at green, and is very low in the violet portion of the spectrum.

There is a cut-off frequency for which the equation above cannot be satisfied. Since the refractive index is a function of frequency (and hence wavelength), the intensity does not continue increasing at ever shorter wavelengths even for ultra-relativistic particles (where v/c approaches 1). At X-ray frequencies, the refractive index becomes less than unity (note that in media the phase velocity may exceed c without violating relativity) and hence no X-ray emission (or shorter wavelength emissions such as gamma rays) would be observed. However, X-rays can be generated at special frequencies just below those corresponding to core electronic transitions in a material, as the index of refraction is often greater than 1 just below a resonance frequency.

As in sonic booms and bow shocks, the angle of the shock cone is directly related to the velocity of the disruption. The Cherenkov angle is zero at the threshold velocity for the emission of Cherenkov radiation. The angle takes on a maximum as the particle speed approaches the speed of light. Hence, observed angles of incidence can be used to compute the direction and speed of a Cherenkov radiation-producing charge.

Cherenkov radiation can be generated in the eye by charged particles hitting the vitreous humour, giving the impression of flashes.

Ilya Frank

Ilya Mikhailovich Frank was a Soviet and winner of the Nobel Prize for Physics in 1958 jointly with Pavel Alekseyevich Cherenkov and Igor Y. Tamm, also of the Soviet Union. He received the award for his work in explaining the phenomenon of Cherenkov radiation.

Born	23 October 1908, St. Petersburg, Russia
Died	22 June 1990 (age 81) Moscow, Russia
Fields	Nuclear physics
Institutions	Moscow State University
Alma mater	Moscow State University
Known for	Cerenkov radiation
Notable awards	Nobel Prize in Physics (1958)

CHERENKOV RADIATION (1958)

While relativity holds that the speed of light *in a vacuum* is a universal constant (c), the speed at which light propagates in a material may be significantly less than c. For example, the speed of the propagation of light in water is only $0.75c$. Matter can be accelerated beyond this speed during nuclear reactions and in particle accelerators. Cherenkov radiation results when a charged particle, most commonly an electron, travels through a dielectric medium with a speed greater than that at which light would otherwise propagate in the same medium.

Moreover, the velocity that must be exceeded is the phase velocity of light rather than the group velocity of light. The phase velocity can be altered dramatically by employing a periodic medium, and in that case one can even achieve Cherenkov radiation with *no* minimum particle velocity, a phenomenon known as the Smith-Purcell effect. In a more complex periodic medium, such as a photonic crystal, one can also obtain a variety of other anomalous Cherenkov effects, such as radiation in a backwards direction Cherenkov radiation forms an acute angle with the particle.

As a charged particle travels, it disrupts the local electromagnetic field (EM) in its medium. Electrons in the atoms of the medium will be displaced, and the atoms become polarized by the passing EM field of a charged particle. Photons are emitted as an insulator's electrons restore themselves to equilibrium after the disruption has passed. In normal circumstances, these photons destructively interfere with each other and no radiation is detected. However, when a disruption which travels faster than light is propagating through the medium, the photons constructively interfere and intensify the observed radiation.

A common analogy is the sonic boom of a supersonic aircraft or bullet. The sound waves generated by the supersonic body propagate at the speed of sound itself; as such, the waves cannot propagate away from the body and form a shock front.

In a similar way, a charged particle can generate a photonic shock wave as it travels through an insulator.

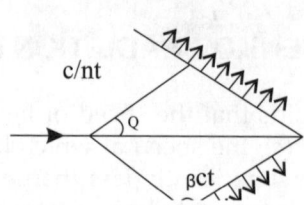

The particle travels in a medium with speed v_p such that $c / n < v_p < c$, where c is speed of light, and n is the refractive index of the medium. We define the ratio between the speed of the particle and the speed of light as $\beta = v_p / c$. The emitted light waves travel at speed $v_{em} = c / n$. The left corner of the triangle represents the location of the superluminal particle at some initial moment ($t=0$). The right corner of the triangle is the location of the particle at some later time t. In the given time t, the particle travels the distance $x_p = v_p t = \beta c t$.

whereas the emitted electromagnetic waves are constricted to travel the distance $x_{em} = v_{em} t = \dfrac{c}{n} t$. So: $\cos\theta = \dfrac{1}{n\beta}$

Igor Tamm

 Igor Yevgenyevich Tamm was a Soviet physicist, mathematician and a Nobel laureate. He was the Nobel Laureate in Physics for the year 1958 together with Pavel Cherenkov and Ilya Frank for the discovery and the interpretation of the Cherenkov-Vavilov effect.

Born	8 July 1895, Vladivostok, Russian Empire
Died	12 April 1971 (age 75), Moscow, Russian SFSR, Soviet Union
Nationality	Soviet Union
Fields	Physics
Known for	Cherenkov-Vavilov effect
Notable awards	Nobel Prize in Physics (1958)

In 1951, together with Andrei Sakharov, Tamm proposed a tokamak system of the realization of CTF on the basis of toroidal magnetic thermonuclear reactor and soon after the first such devices were built by the INF, resulting the T-3 Soviet magnetic confinement device from 1968, when the plasma parameters unique for that time were obtained, of showing the temperatures in their machine to be over an order of magnitude higher than what was expected by the rest of the community. The western scientists visited the experiment and verified the high temperatures and confinement, sparking a wave of optimism for the prospects of the tokamak as well as construction of new experiments, which is still the dominant magnetic confinement device today.

CHERENKOV-VAVILOV EFFECT (1958)

Cherenkov radiation is commonly used in experimental particle physics for particle identification. One could measure (or put limits on) the velocity of an electrically charged elementary particle by the properties of the Cherenkov light it emits in a certain medium. If the momentum of the particle is measured independently, one could compute the mass of the particle by its momentum and velocity, and hence identify the particle.

The simplest type of particle identification device based on a Cherenkov radiation technique is the threshold counter, which gives an answer as to whether the velocity of a charged particle is lower or higher than a certain value ($v_0 = c / n$, where c is the speed of light, and n is the refractive index of the medium) by looking at whether this particle does or does not emit Cherenkov light in a certain medium. Knowing particle momentum, one can separate particles lighter than a certain threshold from those heavier than the threshold.

The most advanced type of a detector is the RICH, or ring imaging Cherenkov detector, developed in the 1980s. In a RICH detector, a cone of Cherenkov light is produced when a high speed charged particle traverses a suitable gaseous or liquid medium, often called radiator. This light cone is detected on a position sensitive planar photon detector, which allows reconstructing a ring or disc, the radius of which is a measure for the Cherenkov emission angle. Both focusing and proximity-focusing detectors are in use. In a focusing RICH detector, the photons are collected by a spherical mirror and focused onto the photon detector placed at the focal plane. The result is a circle with a radius independent of the emission point along the particle track. This scheme is suitable for low refractive index radiators, i.e. gases, due to the larger radiator length needed to create enough photons. In the more compact proximity-focusing design, a thin radiator volume emits a cone of Cherenkov light which traverses a small distance–the proximity gap, and is detected on the photon detector plane. The image is a ring of light, the radius of which is defined by the Cherenkov emission angle and the proximity gap. The ring thickness is determined by the thickness of the radiator. An example of a proximity gap RICH detector is the High Momentum Particle Identification (HMPID), a detector currently under construction for ALICE (A Large Ion Collider Experiment), one of the six experiments at the LHC (Large Hadron Collider) at CERN.

Emilio G. Segrè

 Emilio Gino Segrè was an Italian physicist and Nobel laureate in physics, who with Owen Chamberlain, discovered antiprotons, a sub-atomic antiparticle.

Born	1 February 1905, Tivoli, Italy
Died	22 April 1989 (age 84)
Institutions	Los Alamos National Laboratory, University of California, Berkeley, University of Palermo, University of Rome La Sapienza
Alma mater	University of Rome La Sapienza
Doctoral advisor	Enrico Fermi
Known for	Discovery of the antiproton, Discovery of technetium, Discovery of astatine
Notable awards	Nobel Prize in Physics (1959)

ANTIPROTON (1959)

The antiproton is the antiparticle of the proton. Antiprotons are stable, but they are typically short-lived since any collision with a proton will cause both particles to be annihilated in a burst of energy.

The existence of the antiproton with –1 electric charge, opposite to the +1 electric charge of the proton, was predicted by Paul Dirac in his 1933 Nobel Prize lecture. Dirac received the Nobel Prize for his previous 1928 publication of his Dirac Equation that predicted the existence of + and – solutions to the Energy Equation ($E = mc^2$) of Einstein and the existence of the antimatter positive charge electron, the spin opposite of the negative charge electron.

The antiproton was experimentally confirmed in 1955 by University of California, Berkeley physicists Emilio Segrè and Owen Chamberlain, for which they were awarded the 1959 Nobel Prize in Physics. An antiproton consists of two up antiquark and one down antiquark. The properties of the antiproton that have been measured all match the corresponding properties of the proton, with the exception that the proton has opposite electric charge and magnetic moment than the proton. The question of ways matter is different from antimatter remains an open problem, in order to explain how our universe survived the Big Bang and why so little remains of antimatter today in our solar system.

Technetium: Technetium is the chemical element with atomic number 43 and symbol Tc. It is the lowest atomic number element without any stable isotopes. Nearly all of technetium is produced synthetically and only minute amounts are found in nature. Naturally occurring technetium occurs as a spontaneous fission product in uranium ore or by neutron capture in molybdenum ores. The chemical properties of this silvery gray, crystalline transition metal are intermediate between rhenium and manganese.

Many of technetium's properties were predicted by Dmitri Mendeleev before the element was discovered. Mendeleev noted a gap in his periodic table and gave the undiscovered element the provisional name *ekamanganese*. In 1937 technetium became the first predominantly artificial element to be produced, hence its name.

Astatine: Astatine is a radioactive chemical element with the symbol At and atomic number 85. It is the second-heaviest of the discovered halogens. Although astatine is produced by radioactive decay in nature, due to its short half life, it is found only in minute amounts. Astatine was first produced by Dale R. Corson, Kenneth Ross MacKenzie, and Emilio Segrè in 1940. Three years passed before traces of astatine were also found in natural minerals. Until recently most of the physical and chemical characteristics of astatine were inferred from comparison with other elements. Some astatine isotopes are used as alpha-particle emitters in science applications, and medical applications for astatine 211 have been tested. Astatine is currently the rarest naturally occurring element, with less than 30g estimated to be contained in the entire Earth's crust.

Owen Chamberlain

Owen Chamberlain was an American physicist, and Nobel laureate in physics for his discovery, with collaborator Emilio Segrè, of antiprotons, a sub-atomic antiparticle.

Born	10 July 1920, San Francisco, California, USA
Died	28 February 2006 (age 85), Berkeley, California, USA
Nationality	United States
Fields	Physics
Institutions	Los Alamos National Laboratory
Alma mater	Dartmouth College, University of California, Berkeley, University of Chicago
Known for	Particle physics
Notable awards	Nobel Prize in Physics (1959)

ANTIPARTICLE (1959)

Corresponding to most kinds of particles, there is an associated antiparticle with the same mass and opposite electric charge. For example, the antiparticle of the electron is the positively charged antielectron, or positron, which is produced naturally in certain types of radioactive decay.

The laws of nature are very nearly symmetrical with respect to particles and antiparticles. For example, an antiproton and a positron can form an antihydrogen atom, which has almost exactly the same properties as a hydrogen atom. A physicist whose body was made of antimatter, doing experiments in a laboratory also made of antimatter, using chemicals and substances made of antiparticles, would find almost exactly the same results in all experiments. This leads to the question of why the formation of matter after the Big Bang resulted in a universe consisting almost entirely of matter, rather than being a half-and-half mixture of matter and antimatter. The discovery of CP violation helped to shed light on this problem by showing that this symmetry, originally thought to be perfect, was only approximate.

Particle-antiparticle pairs can annihilate each other, producing photons; since the charges of the particle and antiparticle are opposite, charge is conserved. For example, the antielectrons produced in natural radioactive decay quickly annihilate themselves with electrons, producing pairs of gamma rays.

Antiparticles are produced naturally in beta decay, and in the interaction of cosmic rays in the Earth's atmosphere. Because charge is conserved, it is not possible to create an antiparticle without either destroying a particle of the same charge (as in beta decay), or creating a particle of the opposite charge. The latter is seen in many processes in which both a particle and its antiparticle are created simultaneously, as in particle accelerators. This is the inverse of the particle-antiparticle annihilation process.

Although particles and their antiparticles have opposite charges, electrically neutral particles need not be identical to their antiparticles. The neutron, for example, is made out of quarks, the antineutron from antiquarks, and they are distinguishable from one another because neutrons and antineutrons annihilate each other upon contact. However, other neutral particles are their own antiparticles, such as photons, the hypothetical gravitons, and WIMPs. These are called Majorana particles and can annihilate with themselves.

In 1932, soon after the prediction of positrons by Paul Dirac, Carl D. Anderson found that cosmic-ray collisions produced these particles in a cloud chamber, a particle detector in which moving electrons (or positrons) leave behind trails as they move through the gas. The electric charge-to-mass ratio of a particle can be measured by observing the curling of its cloud-chamber track in a magnetic field. Originally, positrons, because of the direction that their paths curled, were mistaken for electrons travelling in the opposite direction.

Donald A. Glaser

Donald Arthur Glaser is an American physicist, neurobiologist, and Nobel Prize in Physics laureate for his invention of the Bubble chamber used in subatomic particle physics.

Born	21 September 1926 (age 83), Cleveland, Ohio
Fields	Physics
Institutions	University of Michigan, University of California at Berkeley
Alma mater	Case Western Reserve University, California Institute of Technology
Doctoral advisor	Carl David Anderson
Known for	Invention of bubble chamber
Notable awards	Nobel Prize in Physics (1960), Elliott Cresson Medal (1961)

BUBBLE CHAMBER (1960)

A bubble chamber is a vessel filled with a superheated transparent liquid (most often liquid hydrogen) used to detect electrically charged particles moving through it. It was invented in 1952 by Donald A. Glaser, for which he was awarded the 1960 Nobel Prize in Physics. Anecdotally, Glaser was inspired by the bubbles in a glass of beer; however, in a 2006 talk, he refuted this story, saying that although beer was not the *inspiration* for the bubble chamber, he did experiments using beer to fill early prototypes.

Cloud chambers work on the same principles as bubble chambers, only they are based on super cooled gas rather than superheated liquid. While bubble chambers were extensively used in the past, they have now mostly been supplanted by wire chambers and spark chambers. Notable bubble chambers include the Big European Bubble Chamber (BEBC) and Gargamelle.

The bubble chamber is similar to a cloud chamber in application and basic principle. It is normally made by filling a large cylinder with a liquid heated to just below its boiling point. As particles enter the chamber, a piston suddenly decreases its pressure, and the liquid enters into a superheated, metastable phase. Charged particles create an ionization track, around which the liquid vaporizes, forming microscopic bubbles. Bubble density around a track is proportional to a particle's energy loss.

Bubbles grow in size as the chamber expands, until they are large enough to be seen or photographed. Several cameras are mounted around it, allowing a three-dimensional image of an event to be captured. Bubble chambers with resolutions down to a few μm have been operated.

The entire chamber is subject to a constant magnetic field, which causes charged particles to travel in helical paths whose radius is determined by their charge-to-mass ratios. Since the magnitude of the charge of all known charged, long-lived subatomic particles is the same as that of an electron, their radius of curvature must be proportional to their momentum. Thus, by measuring their radius of curvature, their momentum can be determined.

Notable discoveries made by bubble chamber include the discovery of weak neutral currents at Gargamelle in 1973, which establish the soundness of the electroweak theory and paved the way to the discovery of the W and Z bosons in 1983 (at the UA1 and UA2 experiments). Recently, bubble chambers have been used in research on WIMPs, at COUPP and PICASSO.

Robert Hofstadter

Robert Hofstadter was an American physicist. He was the winner of the 1961 Nobel Prize in Physics for his pioneering studies of electron scattering in atomic nuclei and for his consequent discoveries concerning the structure of nucleons.

Born	5 February 1915, New York City, New York
Died	17 November 1990 (age 75)
Nationality	United States
Fields	Physics
Institutions	Stanford University
Alma mater	City College of New York, Princeton University
Known for	Electron scattering, Atomic nuclei
Notable awards	Nobel Prize in Physics (1961)

NUCLEON

In physics, a nucleon is a collective name for two particles: the neutron and the proton. These are the two constituents of the atomic nucleus. Until the 1960s, the nucleons were thought to be elementary particles. Now they are known to be composite particles, each made of three quarks bound together by the so-called strong interaction.

The interaction between two or more nucleons is called internucleon interactions or nuclear force, which is also ultimately caused by the strong interaction.

Nucleons sit at the boundary where particle physics and nuclear physics overlap. Particle physics, particularly quantum chromodynamics, provides the fundamental equations that explain the properties of quarks and of the strong interaction. These equations explain quantitatively how quarks can bind together into protons and neutrons. However, when multiple nucleons are assembled into an atomic nucleus, these fundamental equations become too difficult to solve directly. Instead, nuclides are studied within nuclear physics, which studies nucleons and their interactions by approximations and models, such as the nuclear shell model. These models can successfully explain nuclide properties, for example, whether or not a certain nuclide undergoes radioactive decay.

The proton and neutron are both baryons and both fermions. In the terminology of particle physics, these two particles make up an isospin doublet (I = 1/2). This explains why their masses are so similar, with the neutron just 0.1% heavier than the proton.

A neutron by itself is an unstable particle: It undergoes β-decay (a type of radioactive decay) by turning into a proton, electron, and electron antineutrino, with a half-life around ten minutes. A proton by itself is thought to be stable, or at least its lifetime is too long to measure.

Inside a nucleus, on the other hand, both protons and neutrons can be stable or unstable, depending on the nuclide. Inside some nuclides, a neutron can turn into a proton as described above; inside other nuclides the reverse can happen, where a proton turns into a neutron through β+ decay or electron capture; and inside still other nuclides, both protons and neutrons are stable and do not change form.

ELECTRON SCATTERING (1961)

Electron scattering is the process whereby an electron is deflected from its original trajectory.

Electrons are charged particles and are acted upon by the electromagnetic forces. They are scattered by other charged particles through the electrostatic Coulomb forces. Furthermore, if a magnetic field is present, a traveling electron will be deflected by the Lorentz force. An extremely accurate description of all electron scattering, including quantum and relativistic aspects, is given by the theory of quantum electrodynamics.

Rudolf Mössbauer

Rudolf Ludwig Mössbauer is a German physicist who studied gamma rays from nuclear transitions. Mössbauer was born in Munich, where he also studied physics at the Technical University of Munich (TUM) and did his PhD with Heinz Maier-Leibnitz.

Along with Robert Hofstadter of the United States he won the Nobel Prize in Physics in 1961 for his 1957 discovery of the Mössbauer effect research which he carried out as a PhD student at the Institute for Physics of the Max Planck Institute for Medical Research in Heidelberg.

Born	31 January 1929 (age 81), Munich, Germany
Fields	Nuclear and atomic physics
Institutions	Technical University of Munich, Caltech
Alma mater	Technical University of Munich
Doctoral advisor	Heinz Maier-Leibnitz
Known for	Mössbauer effect, Mössbauer spectroscopy
Notable awards	Nobel Prize in Physics (1961), Elliott Cresson Medal (1961)

MÖSSBAUER EFFECT (1961)

The Mössbauer effect is a physical phenomenon discovered by the German physicist Rudolf Mössbauer in 1957. It involves the resonant and recoil-free emission and absorption of gamma ray photons by atoms bound in a solid form and forms the basis of Mössbauer spectroscopy.

The emission and absorption of X-rays by gases had been observed previously, and it was expected that a similar phenomenon would be found for gamma rays, which are created by nuclear transitions (as opposed to X-rays, which are typically produced by electronic transitions). However, attempts to observe gamma-ray resonance in gases failed due to energy being lost to recoil, preventing resonance (the Doppler effect also broadens the gamma-ray spectrum). Mössbauer was able to observe resonance in solid iridium, which raised the question of why gamma-ray resonance was possible in solids, but not in gases. Mössbauer proposed that, for the case of atoms bound into a solid, under certain circumstances a fraction of the nuclear events could occur essentially without recoil. He attributed the observed resonance to this recoil-free fraction of nuclear events. This discovery was rewarded with the Nobel Prize in Physics in 1961 together with Robert Hofstadter's research of electron scattering in atomic nuclei.

MÖSSBAUER SPECTROSCOPY

Mössbauer spectroscopy is a spectroscopic technique based on the recoil-free, resonant absorption and emission of gamma-rays in solids. This resonant emission and absorption was first observed by Rudolf Mössbauer during his graduate studies in 1957, and is called the Mössbauer effect in his honor. Mössbauer received a Nobel Prize in 1961 for this work.

Like NMR spectroscopy, Mössbauer spectroscopy probes tiny changes in the energy levels of an atomic nucleus in response to its environment. Typically, three types of nuclear interaction may be observed: an isomer shift, also known as a chemical shift; quadrupole splitting; and, magnetic or hyperfine splitting, also known as the Zeeman effect. Due to the high energy and extremely narrow line widths of gamma rays, Mössbauer spectroscopy is one of the most sensitive techniques in terms of energy (and hence frequency) resolution, capable of detecting change in just a few parts per 10

Lev Landau

Lev Davidovich Landau was a prominent Soviet physicist who made fundamental contributions to many areas of theoretical physics.

His accomplishments include the co-discovery of the density matrix method in quantum mechanics, the quantum mechanical theory of diamagnetism, the theory of superfluidity, the theory of second order phase transitions, the Ginzburg-Landau theory of superconductivity, the explanation of Landau damping in plasma physics, the Landau pole in quantum electrodynamics, and the two-component theory of neutrinos. He received the 1962 Nobel Prize in Physics for his development of a mathematical theory of superfluidity that accounts for the properties of liquid helium II at a temperature below 2.17K (−270.98°C).

Born	22 January 1908, Baku, Azerbaijan, Russian Empire
Died	1 April 1968 (age 60), Moscow, Soviet Union
Residence	Soviet Union
Citizenship	Soviet Union
Fields	Theoretical Physics
Institutions	Baku State University, Kharkiv University, Kharkiv Polytechnic Institute, Institute for Physical Problems, MSU Faculty of Physics
Alma mater	Leningrad State, University Leningrad Physico-Technical Institute
Doctoral advisor	Alexei Alexeyevich Abrikosov, Isaak Markovich Khalatnikov
Notable students	Evgeny Lifshitz
Known for	Superfluidity, superconductivity, course of theoretical physics
Notable awards	Nobel Prize in Physics (1962)

SUPER FLUID (1962)

Super fluidity is a phase of matter or description of heat capacity in which unusual effects are observed when liquids, typically of helium 4 or helium 3, overcome friction by surface interaction when at a stage at which the liquid's viscosity becomes zero. Also known as a major facet in the study of quantum hydrodynamics, it was discovered by Pyotr Kapitsa, John F. Allen, and Don Misener in 1937 and has been described through phenomenological and microscopic theories. In the 1950s Hall and Vinen performed experiments establishing the existence of quantized vortex lines. In the 1960s, Rayfield and Reif established the existence of quantized vortex rings. Packard has observed the intersection of vortex lines with the free surface of the fluid, and Avenel and Varoquaux have studied the Josephson effect in super fluid ^{4}He.

L. D. Landau's phenomenological and semi-microscopic theory of super fluidity of ^{4}He earned him the Nobel Prize in Physics in 1962. Assuming that sound waves are the most important excitations in ^{4}He at low temperatures, he showed that ^{4}He flowing past a wall would not spontaneously create excitations if the flow velocity was less than the sound velocity. In this model, the sound velocity is the critical velocity above which super fluidity is destroyed.

Landau also showed that the sound wave and other excitations could equilibrate with one another and flow separately from the rest of the ^{4}He called the condensate.

From the momentum and flow velocity of the excitations he could then define a normal fluid density, which is zero at zero temperature and increases with temperature. At the so-called Lambda temperature, where the normal fluid density equals the total density, the ^{4}He is no longer super fluid.

To explain the early specific heat data on super fluid ^{4}He, Landau posited the existence of a type of excitation he called a roton, but as better data became available he considered that the roton was the same as a high momentum version of sound.

Eugene Wigner

Eugene Paul "E. P." Wigner was a Hungarian American physicist and mathematician. He received the Nobel Prize in Physics in 1963 for his contributions to the theory of the atomic nucleus and the elementary particles, particularly through the discovery and application of fundamental symmetry principles.

Born	17 November 1902, Budapest, Austria-Hungary
Died	1 January 1995 (age 92), Princeton, New Jersey, United States
Residence	USA
Citizenship	American (post-1937), Hungarian (pre-1937)
Fields	Theoretical Physics, Atomic Physics, Nuclear Physics, Solid State Physics
Institutions	University of Göttingen, University of Wisconsin–Madison, Princeton University, Manhattan project
Alma mater	Technische Hochschule Berlin
Doctoral advisor	Michael Polanyi, academic advisors: László Rátz, Richard Becker
Doctoral students	John Bardeen, Victor Frederick Weisskopf, Marcos Moshinsky, Abner Shimony, Edwin Thompson Jaynes, Frederick Seitz, Conyers Herring, Jack H. Irving, Frederick Tappert
Known for	Law of conservation of parity, Wigner D-matrix, Wigner-Eckart theorem, Wigner's friend, Wigner semicircle distribution, Wigner's classification, Wigner quasi-probability distribution, Wigner crystal, Wigner effect, Wigner-Seitz cell, Relativistic Breit-Wigner distribution, modified Wigner distribution function, Wigner-d'Espagnat inequality, Gabor-Wigner transform, Wigner's theorem, Wigner distribution, Jordan-Wigner transformation, Newton-Wigner localization, Wigner-Seitz radius, 6-j symbol, 9-j symbol
Notable awards	Enrico Fermi Award (1958), Max Planck Medal (1961), Nobel Prize in Physics (1963), National Medal of Science (1969)

WIGNER EFFECT (1963)

The Wigner effect, also known as the discomposition effect, is the displacement of atoms in a solid caused by neutron radiation. Any solid can be affected by the Wigner effect, but the effect is of most concern in neutron moderators, such as graphite, that are used to slow down fast neutrons. The material surrounding the moderator receives a much smaller amount of neutron radiation, and from slower neutrons, and is not as worrisome. An interstitial atom and its associated vacancy are known as a Frenkel defect. To create the Wigner effect, neutrons that collide with the atoms in a crystal structure must have enough energy to displace them from the lattice. This amount is approximately 25 eV. A neutron's energy can vary widely but it is not uncommon to have energies up to and exceeding 10 Mev in the center of a nuclear reactor. A neutron with a significant amount of energy will create a displacement cascade in a matrix via elastic collisions. For example, a 1 Mev neutron striking graphite will create 900 displacements, however not all displacements will create defects because some of the struck atoms will find and fill the vacancies that were either small pre-existing voids or vacancies newly formed by the other struck atoms. The atoms that do not find a vacancy come to rest in non-ideal locations; that is, not along the symmetrical lines of the lattice. These atoms are referred to as interstitial atoms, or simply interstitials. Because these atoms are not in the ideal location they have an energy associated with them, much like a ball at the top of a hill has gravitational potential energy. When large amounts of interstitials have accumulated they pose a risk of releasing all of their energy suddenly, creating a temperature spike. Sudden unplanned increases in temperature can present a large risk for certain types of nuclear reactors with low operating temperatures and were the indirect cause of the Windscale fire. Accumulation of energy in irradiated graphite has been recorded as high as 2.7 kJ/g, but is typically much lower than this. Despite some reports, Wigner energy buildup had nothing to do with the Chernobyl disaster: This reactor, like all contemporary power reactors, operated at a high enough temperature to allow the displaced graphite structure to realign itself before any potential energy could be stored.

Maria Goeppert-Mayer

Maria Goeppert-Mayer was a German born American theoretical physicist, and Nobel laureate in Physics for proposing the nuclear shell model of the atomic nucleus. She is the second female laureate in physics, after Marie Curie.

It was during her time at Chicago and Argonne that she developed a mathematical model for the structure of nuclear shells, the work for which she was awarded the Nobel Prize in Physics in 1963, shared with J. Hans D. Jensen and Eugene Paul Wigner.

Born	28 June 1906, Kattowitz, German Empire
Died	20 February 1972 (age 65), San Diego, California, United States
Nationality	Germany
Citizenship	United States
Fields	Physics
Institutions	Los Alamos Laboratory, Argonne National Laboratory
Alma mater	University of Göttingen
Doctoral advisor	Max Born
Known for	Nuclear shell structure
Notable awards	Nobel Prize in Physics (1963)

NUCLEAR SHELL MODEL (1963)

In nuclear physics and nuclear chemistry, the nuclear shell model is a model of the atomic nucleus which uses the Pauli exclusion principle to describe the structure of the nucleus in terms of energy levels. The model was developed in 1949 following independent work by several physicists, most notably Eugene Paul Wigner, Maria Goeppert-Mayer and J. Hans D. Jensen, who shared the 1963 Nobel Prize in Physics for their contributions.

The shell model is partly analogous to the atomic shell model which describes the arrangement of electrons in an atom, in that a filled shell results in greater stability. When adding nucleons (protons or neutrons) to a nucleus, there are certain points where the binding energy of the next nucleon is significantly less than the last one. This observation, that there are certain magic numbers of nucleons: 2, 8, 20, 28, 50, 82, 126 which are more tightly bound than the next higher number, is the origin of the shell model.

The shells exist for both protons and neutrons individually, so that we can speak of magic nuclei where one nucleon type is at a magic number, and doubly magic nuclei, where both are. Due to some variations in orbital filling, the upper magic numbers are 126 and, speculatively, 184 for neutrons but only 114 for protons. This has a relevant role in the search of the so-called island of stability. Besides, there have been found some semimagic numbers, noticeably $Z = 40$.

In order to get these numbers, the nuclear shell model starts from an average potential with a shape something between the square well and the harmonic oscillator. To this potential a spin orbit term is added. Even so, the total perturbation does not coincide with experiment, and an empirical spin orbit coupling, named the Nilsson Term, must be added with at least two or three different values of its coupling constant, depending on the nuclei being studied.

Nevertheless, the magic numbers of nucleons, as well as other properties, can be arrived at by approximating the model with a three-dimensional harmonic oscillator plus a spin-orbit interaction. A more realistic but also complicated potential is known as Woods Saxon potential.

J. Hans D. Jensen

Johannes Hans Daniel Jensen was a German nuclear physicist. During World War II, he worked on the German nuclear energy project, also known as the Uranium Club, in which he made contributions to the separation of uranium isotopes.

After the war, he was a professor at the University of Heidelberg. He was a visiting professor at the University of Wisconsin-Madison, the Institute for Advanced Study, Indiana University, and the California Institute of Technology. Jensen shared half of the 1963 Nobel Prize for Physics with Maria Göppert-Mayer for their proposal of the nuclear shell model.

Born	25 June 1907, Hamburg
Died	11 February 1973, Heidelberg
Nationality	German
Fields	Physics
Alma mater	University of Hamburg
Doctoral advisor	Wilhelm Lenz
Notable awards	Nobel Prize in Physics (1963)

PROPERTIES OF NUCLEI (1963)

This model also predicts or explains with some success other properties of nuclei, in particular spin and parity of nuclei ground states, and to some extent their excited states as well. Take $^{17}_{8}O^{9}$ as an example, its nucleus has eight protons filling the two first proton shells, eight neutrons filling the two first neutron shells, and one extra neutron. All protons in a complete proton shell have total angular momentum zero, since their angular momenta cancel each other; The same is true for neutrons. All protons in the same level (n) have the same parity (either +1 or −1), and since the parity of a pair of particles is the product of their parities, an even number of protons from the same level (n) will have +1 parity. Thus the total angular momentum of the eight protons and the first eight neutrons is zero, and their total parity is +1. This means that the spin (i.e. angular momentum) of the nucleus, as well as its parity, are fully determined by that of the ninth neutron. This one is in the first (i.e. lowest energy) state of the 3rd shell, and therefore have $n = 2$, giving it +1 parity, and $j = 5/2$. Thus the nucleus of $^{17}_{8}O^{9}$ is expected to have positive parity and spin 5/2, which indeed it has.

For nuclei farther from the magic numbers one must add the assumption that due to the relation between the strong nuclear force and angular momentum, protons or neutrons with the same n tend to form pairs of opposite angular momenta. Therefore a nucleus with an even number of protons and an even number of neutrons has 0 spin and positive parity. A nucleus with an even number of protons and an odd number of neutrons (or vice versa) has the parity of the last neutron (or proton), and the spin equal to the total angular momentum of this neutron (or proton). By last we mean the properties coming from the highest energy level.

In the case of a nucleus with an odd number of protons and an odd number of neutrons, one must consider the total angular momentum and parity of both the last neutron and the last proton. The nucleus parity will be a product of theirs, while the nucleus spin will be one of the possible results of the sum of their angular momenta with other possible results being excited states of the nucleus.

The ordering of angular momentum levels within each shell is according to the principles described above - due to spin-orbit interaction, with high angular momentum states having their energies shifted downwards due to the deformation of the potential (i.e. moving form a harmonic oscillator potential to a more realistic one). For nucleon pairs, however, it is often energetically favorable to be at high angular momentum, even if its energy level for a single nucleon would be higher. This is due to the relation between angular momentum and the strong nuclear force.

Charles Hard Townes

Charles Hard Townes is an American Nobel Prize winning physicist and educator. Townes is known for his work on the theory and application of the maser, on which he got the fundamental patent, and other work in quantum electronics connected with both maser and laser devices. He received the Nobel Prize in Physics in 1964.

Born	28 July 1915 (age 94), Greenville South Carolina
Nationality	United States
Residence	United States
Fields	Physics
Institutions	Bell Labs, Institute for Defense Analyses, Columbia, MIT, Berkeley
Alma mater	Furman University, Duke, Caltech
Doctoral advisor	William Smythe
Doctoral students	James P. Gordon, Robert Boyd, Ali Javan, Raymond Y. Chiao
Known for	Inventing the Maser and Laser
Notable awards	Nobel Prize in Physics (1964), Templeton Prize (2005)

MASER (1964)

A maser is a device that produces coherent electromagnetic waves through amplification by stimulated emission. Historically, maser derives from the original, upper-case acronym MASER (Microwave Amplification by Stimulated Emission of Radiation). The lower-case usage arose from technological development having rendered the original denotation imprecise, because contemporary masers emit EM waves (microwave and radio frequencies) across a broader band of the electromagnetic spectrum; thus, the physicist Charles H. Townes's suggested usage of molecular replacing microwave, for contemporary linguistic accuracy. In 1957, when the optical coherent oscillator was first developed, it was denominated *optical maser*, but usually called laser (Light Amplification by Stimulated Emission of Radiation), the acronym Gordon Gould established in 1957.

Theoretically, the principle of the maser was described by Nikolay Basov and Alexander Prokhorov from Lebedev Institute of Physics at an *All-Union Conference on Radio-Spectroscopy* held by USSR Academy of Sciences in May 1952. They subsequently published their results in October 1954. Independently, Charles H. Townes, J. P. Gordon, and H. J. Zeiger built the first maser at Columbia University in 1953. The device used stimulated emission in a stream of energized ammonia molecules to produce amplification of microwaves at a frequency of 24 gigahertz. Townes later worked with Arthur L. Schawlow to describe the principle of the *optical maser*, or *laser*, which Theodore H. Maiman first demonstrated in 1960. For their research in this field Townes, Basov, and Prokhorov were awarded the Nobel Prize in Physics in 1964.

LASER

Light Amplification by Stimulated Emission of Radiation (LASER or laser) is a mechanism for emitting electromagnetic radiation, typically light or visible light, via the process of stimulated emission. The emitted laser light is (usually) a spatially coherent, narrow low-divergence beam, that can be manipulated with lenses. In laser technology, coherent light denotes a light source that produces (emits) light of in-step waves of identical frequency, phase, and polarization. The laser's beam of coherent light differentiates it from light sources that emit *incoherent* light beams, of random phase varying with time and position. Laser light is generally a narrow-wavelength electromagnetic spectrum monochromatic light; yet, there are lasers that emit a broad spectrum of light, or emit different wavelengths of light simultaneously.

Nikolay Basov

Nikolay Gennadiyevich Basov was a Soviet physicist and educator. For his fundamental work in the field of quantum electronics that led to the development of laser and maser, Basov shared the 1964 Nobel Prize in Physics with Alexander Prokhorov and Charles Hard Townes.

Born	14 December 1922, Usman, Russia
Died	1 July 2001 (age 78)
Fields	Physics
Institutions	Lebedev Physical Institute
Alma mater	Moscow Engineering Physics Institute
Known for	Invention of lasers and masers
Notable awards	Nobel Prize in Physics (1964)

LASER PHYSICS (1964)

The gain medium of a laser is a material of controlled purity, size, concentration, and shape, which amplifies the beam by the process of stimulated emission. It can be of any state: gas, liquid, solid or plasma. The gain medium absorbs pump energy, which raises some electrons into higher-energy (excited) quantum states. Particles can interact with light by either absorbing or emitting photons. Emission can be spontaneous or stimulated. In the latter case, the photon is emitted in the same direction as the light that is passing by. When the number of particles in one excited state exceeds the number of particles in some lower-energy state, population inversion is achieved and the amount of stimulated emission due to light that passes through is larger than the amount of absorption. Hence, the light is amplified. By itself, this makes an optical amplifier. When an optical amplifier is placed inside a resonant optical cavity, one obtains a laser.

The light generated by stimulated emission is very similar to the input signal in terms of wavelength, phase, and polarization. This gives laser light its characteristic coherence, and allows it to maintain the uniform polarization and often monochromaticity established by the optical cavity design.

The optical cavity, a type of cavity resonator, contains a coherent beam of light between reflective surfaces so that the light passes through the gain medium more than once before it is emitted from the output aperture or lost to diffraction or absorption. As light circulates through the cavity, passing through the gain medium, if the gain (amplification) in the medium is stronger than the resonator losses, the power of the circulating light can rise exponentially. But each stimulated emission event returns a particle from its excited state to the ground state, reducing the capacity of the gain medium for further amplification. When this effect becomes strong, the gain is said to be *saturated*. The balance of pump power against gain saturation and cavity losses produces an equilibrium value of the laser power inside the cavity; this equilibrium determines the operating point of the laser. If the chosen pump power is too small, the gain is not sufficient to overcome the resonator losses, and the laser will emit only very small light powers. The minimum pump power needed to begin laser action is called the *lasing threshold*. The gain medium will amplify any photons passing through it, regardless of direction; but only the photons aligned with the cavity manage to pass more than once through the medium and so have significant amplification.

Laser science is principally concerned with quantum electronics, laser construction, optical cavity design, the physics of producing a population inversion in laser media, and the temporal evolution of the light field in the laser. It is also concerned with the physics of laser beam propagation, particularly the physics of Gaussian beams, with laser applications, and with associated fields such as nonlinear optics and quantum optics.

Alexander Prokhorov

Alexander Mikhaylovich Prokhorov was a Russian physicist known for his pioneering research on lasers and masers for which he was awarded Nobel Prize in Physics in 1964.

Born	11 July 1916, Atherton, Queensland, Australia
Died	8 January 2000 (age 83), Moscow, Russia
Nationality	Russian
Fields	Physics
Known for	Lasers and masers
Notable awards	Nobel Prize in Physics (1964)

LASER (1964)

The beam in the cavity and the output beam of the laser, if they occur in free space rather than waveguides (as in an optical fiber laser), are, at best, low order Gaussian beams. However this is rarely the case with powerful lasers. If the beam is not a low-order Gaussian shape, the transverse modes of the beam can be described as a superposition of Hermite-Gaussian or Laguerre-Gaussian beams (for stable-cavity lasers). Unstable laser resonators on the other hand, have been shown to produce fractal shaped beams. The beam may be highly *collimated*, that is being parallel without diverging. However, a perfectly collimated beam cannot be created, due to diffraction. The beam remains collimated over a distance which varies with the square of the beam diameter, and eventually diverges at an angle which varies inversely with the beam diameter. Thus, a beam generated by a small laboratory laser such as a helium-neon laser spreads to about 1.6 kilometers (1 mile) diameter if shone from the Earth to the Moon. By comparison, the output of a typical semiconductor laser, due to its small diameter, diverges almost as soon as it leaves the aperture, at an angle of anything up to 50°. However, such a divergent beam can be transformed into a collimated beam by means of a lens. In contrast, the light from non-laser light sources cannot be collimated by optics as well.

Although the laser phenomenon was discovered with the help of quantum physics, it is not essentially more quantum mechanical than other light sources. The operation of a free electron laser can be explained without reference to quantum mechanics.

The output of a laser may be a continuous constant-amplitude output (known as CW or *continuous wave*); or pulsed, by using the techniques of Q-switching, mode locking, or gain-switching. In pulsed operation, much higher peak powers can be achieved.

Some types of lasers, such as *dye lasers* and *vibronic solid-state lasers* can produce light over a broad range of wavelengths; this property makes them suitable for generating extremely short pulses of light, on the order of a few femtoseconds (10^{-15} s).

The word laser started as an acronym for light amplification by stimulated emission of radiation; in modern usage light broadly denotes electromagnetic radiation of any frequency, not only visible light, hence *infrared laser, ultraviolet laser, X-ray laser*, and so on. Because the microwave predecessor of the laser, the maser, was developed first, devices of this sort operating at microwave and radio frequencies are referred to as masers rather than microwave lasers or radio lasers. In the early technical literature, especially at Bell Telephone Laboratories, the laser was called an optical maser; this term is now obsolete.

Sin-Itiro Tomonaga

 Sin-Itiro Tomonaga or Shin'ichiro Tomonaga was a Japanese physicist, influential in the development of quantum electrodynamics, work for which he was jointly awarded the Nobel Prize in Physics in 1965 along with Richard Feynman and Julian Schwinger.

Born	31 March 1906, Tokyo, Japan
Died	8 July 1979 (age 73), Tokyo, Japan
Fields	Theoretical physics
Institutions	Institute for Advanced Study, Tokyo University of Education
Alma mater	Kyoto Imperial University
Known for	Quantum electrodynamics
Notable awards	Nobel Prize in Physics (1965)

QUANTUM ELECTRODYNAMICS (1965)

Quantum electrodynamics (QED) is the relativistic quantum field theory of electrodynamics. It basically describes how light and matter interact and is the first theory where full agreement between quantum mechanics and special relativity is achieved. QED mathematically describes all phenomena involving electrically charged particles interacting by means of exchange of photons and represents the quantum counterpart of classical electrodynamics giving a complete account of matter and light interaction. One of the founding fathers of QED, Richard Feynman, has called it the jewel of physics for its extremely accurate predictions of quantities like the anomalous magnetic moment of the electron, and the Lamb shift of the energy levels of hydrogen. In technical terms, QED can be described as a perturbation theory of the electromagnetic quantum vacuum.

A first formulation of a quantum theory describing radiation and matter interaction is due to Paul Adrien Maurice Dirac that, during '20, firstly was able to compute the coefficient of spontaneous emission of an atom.

Dirac put forward the quantization of the electromagnetic field seen as an ensemble of harmonic oscillators with the introduction of the concept of creation and annihilation operators of particles. In the following years, with the works by Wolfgang Pauli, Eugene Wigner, Pascual Jordan, Werner Heisenberg and an elegant formulation of quantum electrodynamics due to Enrico Fermi, it was clear that, in principle, it would have been possible to do any computation for whatever physical process involving photons and charged particles. Further studies by Felix Bloch with Arnold Nordsieck, and Victor Weisskopf, in 1937 and 1939, actually showed as such computations were trustful just at a first order of perturbation theory, a problem already pointed out by Robert Oppenheimer. At the next orders in the series there were infinities that made such computations meaningless casting serious doubts on the internal consistency of the theory itself. It appeared to exist a fundamental incompatibility between special relativity and quantum mechanics without any possible solution to work this situation out.

Difficulties with the theory increased through the end of 1940. Improvements in microwave technology made it possible to take more precise measurements of the shift of the levels of a hydrogen atom, now known as the Lamb shift and magnetic moment of the electron. These experiments unequivocally exposed discrepancies which the theory was unable to explain.

Julian Schwinger

Julian Seymour Schwinger was an American theoretical physicist. He is best known for his work on the theory of quantum electrodynamics, in particular for developing a relativistically invariant perturbation theory, and for renormalizing QED to one loop order.

Schwinger was jointly awarded the Nobel Prize in Physics in 1965 for his work on quantum electrodynamics (QED), along with Richard Feynman and Shinichiro Tomonaga. Schwinger's awards and honors were numerous even before his Nobel win. They include the first Albert Einstein Award (1951), the U.S. National Medal of Science (1964), honorary D.Sc. degrees from Purdue University (1961) and Harvard University (1962), and the Nature of Light Award of the U.S. National Academy of Sciences (1949).

Born	12 February 1918, New York City, New York, USA
Died	16 July 1994 (age 76), Los Angeles, California, USA
Nationality	United States
Fields	Physics
Institutions	University of California, Berkeley, Purdue University, Massachusetts Institute of Technology, Harvard University, University of California, Los Angeles
Alma mater	City College of New York, Columbia University
Doctoral advisor	Isidor Isaac Rabi
Doctoral students	Roy Glauber, Ben R. Mottelson, Sheldon Lee Glashow, Walter Kohn, Bryce DeWitt, Daniel Kleitman, Sam Edwards, Gordon Baym, Alain Phares, Kenneth A. Johnson
Known for	Quantum electrodynamics
Notable awards	Nobel Prize in Physics (1965)

QUANTUM ELECTRODYNAMICS (1965)

Difficulties become increasingly stronger on the end of '40. Improvements in microwave technology made possible to realize precision measurements, specially for the shift of the levels of hydrogen atom, nowadays known as Lamb shift, and magnetic moment of the electron. These experiments showed unequivocally discrepancies that the theory was unable to explain. A first indication of a possible way out was given by Hans Bethe. In 1947, while he was traveling by train to reach Schenectady from New York, after attending a conference at Shelter Island about these questions where he addressed a talk, he completed a first non-relativistic computation of the shift of the lines of the hydrogen atom as measured by Lamb and Rutherford. Notwithstanding the limitations implied at such a computation, agreement was excellent. The idea was simply to attach infinities to corrections at mass and charge that were actually fixed to a finite value by experiments. As such, infinities get absorbed in that constants and the final result is finite and in good agreement with experiments. This procedure was named renormalization.

Based on this intuition, it was possible to get fully covariant formulations, that were finite at any order in a perturbation series, of quantum electrodynamics thanks to the fundamental papers of Sin-Itiro Tomonaga, Julian Schwinger, Richard Feynman and Freeman Dyson. Sin-Itiro Tomonaga, Julian Schwinger and Richard Feynman were jointly awarded with a Nobel prize in physics in 1965 for these works: Their contributions, and that relevant of Freeman Dyson, were about covariant and gauge invariant formulations of quantum electrodynamics that allow computations of observables at any order of perturbation theory. Feynman's mathematical technique, based on his diagrams, initially seemed very different from the field-theoretic, operator-based approach of Schwinger and Tomonaga, but Freeman Dyson later showed that the two approaches were equivalent.

Richard Feynman

 Richard Phillips Feynman was an American physicist known for his work in the path integral formulation of quantum mechanics, the theory of quantum electrodynamics and the physics of the superfluidity of supercooled liquid helium, as well as in particle physics.

For his contributions to the development of quantum electrodynamics, Feynman, jointly with Julian Schwinger and Sin-Itiro Tomonaga, received the Nobel Prize in Physics in 1965.

Born	11 May 1918, Far Rockaway, Queens, New York, USA
Died	15 February 1988 (age 69), Los Angeles, California, USA
Residence	United States
Nationality	American
Fields	Physics
Institutions	Manhattan Project, Cornell University, California Institute of Technology
Alma mater	Massachusetts Institute of Technology, Princeton University
Doctoral advisor	John Archibald Wheeler
Academic advisors	Manuel Sandoval Vallarta
Doctoral students	F. L. Vernon, Jr. , Willard H. Wells, Al Hibbs, George Zweig, Giovanni Rossi Lomanitz, Thomas Curtright
Known for	Feynman diagrams, Feynman point, Feynman-Kac formula, Wheeler-Feynman absorber theory, Feynman sprinkler, Feynman Long Division Puzzles, Hellmann-Feynman theorem, Feynman slash notation, Feynman parametrization, Sticky bead argument, One-electron universe, Quantum cellular automata
Notable awards	Albert Einstein Award (1954), E. O. Lawrence Award (1962), Nobel Prize in Physics (1965), Oersted Medal (1972), National Medal of Science (1979)

QUANTUM ELECTRODYNAMICS (1965)

Quantum electrodynamics (QED) is the relativistic quantum field theory of electrodynamics. In essence, it describes how light and matter interact and is the first theory where full agreement between quantum mechanics and special relativity is achieved. QED mathematically describes all phenomena involving electrically charged particles interacting by means of exchange of photons and represents the quantum counterpart of classical electrodynamics giving a complete account of matter and light interaction. One of the founding fathers of QED, Richard Feynman, has called it the jewel of physics for its extremely accurate predictions of quantities like the anomalous magnetic moment of the electron, and the Lamb shift of the energy levels of hydrogen.

The first formulation of a quantum theory describing radiation and matter interaction is due to Paul Adrien Maurice Dirac, who, during 1920, was first able to compute the coefficient of spontaneous emission of an atom.

Dirac described the quantization of the electromagnetic field as an ensemble of harmonic oscillators with the introduction of the concept of creation and annihilation operators of particles. In the following years, with contributions from Wolfgang Pauli, Eugene Wigner, Pascual Jordan, Werner Heisenberg and an elegant formulation of quantum electrodynamics due to Enrico Fermi, physicists came to believe that, in principle, it would be possible to perform any computation for any physical process involving photons and charged particles. However, further studies by Felix Bloch with Arnold Nordsieck, and Victor Weisskopf, in 1937 and 1939, revealed that such computations were reliable only at a first order of perturbation theory, a problem already pointed out by Robert Oppenheimer. At higher orders in the series infinities emerged, making such computations meaningless and casting serious doubts on the internal consistency of the theory itself. With no solution for this problem known at the time, it appeared that a fundamental incompatibility existed between special relativity and quantum mechanics.

Alfred Kastler

Alfred Kastler was a French physicist, and Nobel Prize laureate. He won the Nobel Prize in Physics in 1966 for the discovery and development of optical methods for studying Hertzian resonances in atoms.

Born	3 May 1902, Guebwiller
Died	7 January 1984
Nationality	France
Fields	Physics
Notable awards	Nobel Prize in Physics (1966)

OPTICAL PUMPING (1966)

Optical pumping is a process in which light is used to raise (or pump) electrons from a lower energy level in an atom or molecule to a higher one. It is commonly used in laser construction, to pump the active laser medium so as to achieve population inversion. The technique was developed by 1966 Nobel Prize winner Alfred Kastler in the early 1950s.

Optical pumping is also used to cyclically pump electrons bound within an atom or molecule to a well-defined quantum state. For the simplest case of coherent two-level optical pumping of an atomic species containing a single outer-shell electron, this means that the electron is coherently pumped to a single hyperfine sublevel (labeled m_F), which is defined by the polarization of the pump laser along with the quantum selection rules. Upon optical pumping, the atom is said to be *oriented* in a particular m_F sublevel, however due to the cyclic nature of optical pumping the bound electron will actually be undergoing repeated excitation and decay between upper and lower state sublevels. The frequency and polarization of the pump laser determines which m_F sublevel the atom is oriented in.

In practice, completely coherent optical pumping may not occur due to power-broadening of the line width of a transition and undesirable effects such as hyperfine structure trapping and radiation trapping. Therefore, the orientation of the atom depends more generally on the frequency, intensity, polarization, spectral bandwidth of the laser as well as the line width and transition probability of the absorbing transition.

An optical pumping experiment is commonly found in physics undergraduate laboratories, using rubidium gas isotopes and displaying the ability of radiofrequency (MHz) electromagnetic radiation to effectively pump and unpump these isotopes.

Laser pumping is the act of energy transfer from an external source into the gain medium of a laser. The energy is absorbed in the medium, producing excited states in its atoms. When the number of particles in one excited state exceeds the number of particles in the ground state or a less-excited state, population inversion is achieved. In this condition, the mechanism of stimulated emission can take place and the medium can act as a laser or an optical amplifier. The pump power must be higher than the lasing threshold of the laser.

The pump energy is usually provided in the form of light or electric current, but more exotic sources have been used, such as chemical or nuclear reactions.

A laser pumped with an arc lamp or a flash lamp is usually pumped through the lateral wall of the lasing medium, which is often in the form of a crystal rod containing a metallic impurity or a glass tube containing a liquid dye, in a condition known as side-pumping. To use the lamp's energy most efficiently, the lamps and lasing medium are contained in a reflective cavity that will redirect most of the lamp's energy into the rod or dye cell.

Hans Albrecht Bethe

 Hans Albrecht Bethe was a German-American physicist, and Nobel laureate in physics for his work on the theory of stellar nucleosynthesis. A versatile theoretical physicist, Bethe also made important contributions to quantum electrodynamics, nuclear physics, solid-state physics and particle astrophysics.

During World War II, he was head of the Theoretical Division at the secret Los Alamos laboratory developing the first atomic bombs. There he played a key role in calculating the critical mass of the weapons, and did theoretical work on the implosion method used in both the Trinity test and the Fat Man weapon dropped on Nagasaki, Japan. In 1967, Bethe was awarded the Nobel Prize in Physics for his contributions to the theory of nuclear reactions, especially his discoveries concerning the energy production in stars. His postulate was that the source of this stellar nucleosynthesis was thermonuclear reactions in which hydrogen is converted into helium.

Born	2 July 1906, Strasbourg, Germany
Died	6 March 2005 (age 98), Ithaca, New York, US
Residence	United States
Nationality	German, American
Fields	Nuclear Physics
Institutions	University of Tübingen, Cornell University, University of Manchester
Alma mater	University of Frankfurt, University of Munich
Doctoral advisor	Arnold Sommerfeld
Doctoral students	Jeffrey Goldstone, Roman Jackiw, Freeman Dyson, Robert Eugene Marshak, John Irwin, P. S. Epstein
Known for	Atomic Physics
Notable awards	Nobel Prize in Physics (1967)

STELLAR NUCLEOSYNTHESIS (1967)

Stellar nucleosynthesis is the collective term for the nuclear reactions taking place in stars to build the nuclei of the elements heavier than hydrogen. Some small quantity of these reactions also occur on the stellar surface under various circumstances. For the creation of elements during the explosion of a star, the term supernova nucleosynthesis is used.

The processes involved began to be understood early in the 20th century, when it was first realized that the energy released from nuclear reactions accounted for the longevity of the Sun as a source of heat and light. The prime energy producer in the sun is the fusion of hydrogen to helium, which occurs at a minimum temperature of 3 million Kelvin.

In 1920, Arthur Eddington, on the basis of the precise measurements of atoms by F.W. Aston, was the first to suggest that stars obtained their energy from nuclear fusion of hydrogen to form helium. In 1928, George Gamow derived what is now called the Gamow factor, a quantum-mechanical formula that gave the probability of bringing two nuclei sufficiently close for the strong nuclear force to overcome the Coulomb barrier. The Gamow factor was used in the decade that followed by Atkinson and Houtermans and later by Gamow himself and Teller to derive the rate at which nuclear reactions would proceed at the high temperatures believed to exist in stellar interiors.

In 1939, in a paper entitled Energy Production in Stars, Hans Bethe analyzed the different possibilities for reactions by which hydrogen is fused into helium. He selected two processes that he believed to be the sources of energy in stars. The first one, the proton-proton chain, is the dominant energy source in stars with masses up to about the mass of the Sun. The second process, the carbon-nitrogen-oxygen cycle, which was also considered by Carl Friedrich von Weizsäcker in 1938, is most important in more massive stars.

Luis W. Alvarez

Luis W. Alvarez was an American experimental physicist and inventor, who spent nearly all of his long professional career on the faculty of the University of California, Berkeley.

The American Journal of Physics commented, Luis Alvarez was one of the most brilliant and productive experimental physicists of the 20th century. He won the Nobel Prize in Physics in 1968, and received over 40 patents, some of which proved commercially viable.

Born	13 June 1911, San Francisco, California, USA
Died	1 September 1988 (age 77)
Fields	Physics
Institutions	University of California, Berkeley
Alma mater	University of Chicago
Notable awards	Nobel Prize in Physics (1968)

HYDROGEN BUBBLE CHAMBER (1968)

Luis W. Alvarez was a Nobel laureate physicist with a long career as a creative and innovative scientist. Known as Luie to everyone, among his many productive years was a period at the Lawrence Berkeley Laboratory during which he developed the high-energy particle detector known as the hydrogen bubble chamber and pioneered the use of digital computer technology to analyze the photographs that it produced.

Indeed, his citation for the Nobel Prize of 1968 says: For his decisive contributions to elementary particle physics, in particular the discovery of a large number of resonance states, made possible through his development of the technique of using hydrogen bubble chamber and data analysis.

Using many of these same computer based data analysis techniques Alvarez also organized and directed a large and fruitful effort in detecting, and investigation of both cosmic rays and electromagnetic radiation at high altitudes.

Alvarez led a group of other physicists, graduate students, engineers, and technicians known originally as the Alvarez Group and later as Group A. The intellectual work of this group was recorded in three levels. The first and highest level was the publication of articles in the physics journals, primarily Physical Review and Physical Review Letters. The second and less formal level was in printed reports known as University of California Radiation Laboratory reports, or UCRLs. These were printed and distributed by the Lawrence Berkeley Lab (known then as the Lawrence Radiation Laboratory). Among the UCRLs were the theses of the scores of students who got their PhDs in the Alvarez Group. The third and lowest level of formality was what came to be known as Alvarez Group Memos, which were the working papers of the group. They were written by almost anyone physicists, programmers, engineers, technicians, often handwritten particularly because the symbols were difficult to type, though many were more formally typed. They cover a wide variety of subjects and provide a fascinating look at the working level of science during this two decade burst of creativity. Though the memos were informal in the sense of presentation, they had an extremely rigorous system of review as they were read and digested by all of the other group members who might be interested in the subject, and who were definitely not shy in providing critiques of the work. They were a particularly good tool in the training of graduate students, both an opportunity for original scientific writing and a requirement to defend the work against the queries and comments of these knowledgeable readers.

Murray Gell-Mann

Murray Gell-Mann is an American physicist who received the 1969 Nobel Prize in physics for his work on the theory of elementary particles.

Born	15 September 1929 (age 80), Manhattan, New York City, U.S.
Residence	United States
Nationality	American
Fields	Physics
Institutions	Santa Fe Institute, California Institute of Technology, University of New Mexico
Alma mater	Yale University, MIT
Doctoral advisor	Victor Weisskopf
Doctoral students	Kenneth G. Wilson, Sidney Coleman, Rod Crewther, James Hartle, Christopher T. Hill, H. Jay Melosh, Barton Zwiebach, Kenneth Young, Todd Brun
Known for	Elementary particles
Notable awards	Nobel Prize in Physics (1969)

ELEMENTARY PARTICLE (1969)

In particle physics, an elementary particle or fundamental particle is a particle not known to have substructure; that is, not known to be made up of smaller particles. If an elementary particle truly has no substructure, then it is one of the basic building blocks of the universe from which all other particles are made. In the Standard Model, the quarks, leptons, and gauge bosons are elementary particles.

Historically, the hadrons (mesons and baryons such as the proton and neutron) and even whole atoms were once regarded as elementary particles. A central feature in elementary particle theory is the early 20th century idea of quanta, which revolutionized the understanding of electromagnetic radiation and brought about quantum mechanics. For mathematical purposes, elementary particles are normally treated as point particles, although some particle theories such as string theory posit a physical dimension.

All elementary particles are either bosons or fermions (depending on their spin). The spin-statistics theorem identifies the resulting quantum statistics that differentiates fermions from bosons. According to this methodology: particles normally associated with matter are fermions, having half-integer spin; they are divided into twelve flavours. Particles associated with fundamental forces are bosons, having integer spin.

- **Fermions:**
  Quarks: Up, down, charm, strange, top, bottom
  Leptons: Electron neutrino, electron, muon neutrino, muon, tau neutrino, tau
- **Bosons:**
  Gauge bosons: Gluon, W and Z bosons, photon
  Other bosons: Higgs boson, graviton

The Standard Model of particle physics contains 12 flavors of elementary fermions, plus their corresponding antiparticles, as well as elementary bosons that mediate the forces and the still undiscovered Higgs boson. However, the Standard Model is widely considered to be a provisional theory rather than a truly fundamental one, since it is not known if it is compatible with Einstein's general relativity. There are likely to be hypothetical elementary particles not described by the Standard Model, such as the graviton, the particle that would carry the gravitational force or the sparticles, super symmetric partners of the ordinary particles.

Hannes Olof Gösta Alfvén

 Hannes Olof Gösta Alfvén was a Swedish electrical engineer, plasma physicist and winner of the 1970 Nobel Prize in Physics for his work on magnetohydrodynamics (MHD). He described the class of MHD waves now known as Alfvén waves.

He was originally trained as an electrical power engineer and later moved to research and teaching in the fields of plasma physics and electrical engineering. Alfvén made many contributions to plasma physics, including theories describing the behavior of aurorae, the Van Allen radiation belts, the effect of magnetic storms on the Earth's magnetic field, the terrestrial magnetosphere, and the dynamics of plasmas in the Milky Way galaxy.

Born	30 May 1908, Norrköping, Sweden
Died	2 April 1995 (age 86), Djursholm, Sweden
Fields	Plasma physics
Institutions	University of Uppsala, Nobel Institute for Physics, Royal Institute of Technology, University of California, San Diego, University of Maryland, College Park, University of Southern California
Alma mater	University of Uppsala
Doctoral students	Carl-Gunne Falthammar, Bibhas De, Wing-Huen Ip
Known for	Magnetohydrodynamics
Notable awards	Nobel Prize in Physics (1970)

MAGNETOHYDRODYNAMICS (1970)

Magnetohydrodynamics (MHD) (magnetofluiddynamics or hydromagnetics) is the academic discipline which studies the dynamics of electrically conducting fluids. Examples of such fluids include plasmas, liquid metals, and salt water. The word magnetohydrodynamics (MHD) is derived from magneto that means magnetic field, and hydro- meaning liquid, and dynamics meaning movement. The field of MHD was initiated by Hannes Alfvén.

The idea of MHD is that magnetic fields can induce currents in a moving conductive fluid, which create forces on the fluid, and also change the magnetic field itself. The set of equations which describe MHD are a combination of the Navier-Stokes equations of fluid dynamics and Maxwell's equations of electromagnetism. These differential equations have to be solved simultaneously, either analytically or numerically. MHD is a continuum theory and as such it cannot treat kinetic phenomena, i.e. those in which the existence of discrete particles or of a non-thermal velocities distribution are important.

The simplest form of MHD, Ideal MHD, assumes that the fluid has so little resistivity that it can be treated as a perfect conductor. This is the limit of infinite magnetic Reynolds number. In ideal MHD, Lenz's law dictates that the fluid is in a sense tied to the magnetic field lines. To explain, in ideal MHD a small rope-like volume of fluid surrounding a field line will continue to lie along a magnetic field line, even as it is twisted and distorted by fluid flows in the system. The connection between magnetic field lines and fluid in ideal MHD fixes the topology of the magnetic field in the fluid. For example, if a set of magnetic field lines are tied into a knot, then they will remain so as long as the fluid/plasma has negligible resistivity. This difficulty in reconnecting magnetic field lines makes it possible to store energy by moving the fluid or the source of the magnetic field. The energy can then become available if the conditions for ideal MHD break down, allowing magnetic reconnection that releases the stored energy from the magnetic field.

In an imperfectly conducting fluid the magnetic field can generally move through the fluid following a diffusion law with the resistivity of the plasma serving as a diffusion constant.

Louis Eugène Félix Néel

Louis Eugène Félix Néel was a French physicist born in Lyon. He studied at the Lycée du Parc in Lyon and was accepted at the École Normale Supérieure in Paris.

He was corecipient (with the Swedish astrophysicist Hannes Alfvén) of the Nobel Prize for Physics in 1970 for his pioneering studies of the magnetic properties of solids. His contributions to solid state physics have found numerous useful applications, particularly in the development of improved computer memory units. About 1930 he suggested that a new form of magnetic behavior might exist; called antiferromagnetism, as opposed to ferromagnetism. Above a certain temperature (the Néel temperature) this behaviour stops. Néel pointed out (1947) that materials could also exist showing ferrimagnetism. Néel has also given an explanation of the weak magnetism of certain rocks, making possible the study of the history of Earth's magnetic field.

Born	22 November 1904, Lyon, France
Died	17 November 2000 (age 95)
Fields	Solid-state physics
Alma mater	École Normale Supérieure
Notable awards	Nobel Prize in Physics (1970)

MAGNETOHYDRODYNAMICS (1970)

Even in physical systems which are large and conductive enough that simple estimates of the Lundquist number suggest that we can ignore the resistivity, resistivity may still be important: many instabilities exist that can increase the effective resistivity of the plasma by factors of more than a billion. The enhanced resistivity is usually the result of the formation of small scale structure like current sheets or fine scale magnetic turbulence, introducing small spatial scales into the system over which ideal MHD is broken and magnetic diffusion can occur quickly. When this happens, magnetic reconnection may occur in the plasma to release stored magnetic energy as waves, bulk mechanical acceleration of material, particle acceleration, and heat.

Magnetic reconnection in highly conductive systems is important because it concentrates energy in time and space, so that gentle forces applied to a plasma for long periods of time can cause violent explosions and bursts of radiation.

When the fluid cannot be considered as completely conductive, but the other conditions for ideal MHD are satisfied, it is possible to use an extended model called resistive MHD. This includes an extra term in Ampere's law which models the collisional resistivity. Generally MHD computer simulations are at least somewhat resistive because their computational grid introduces a numerical resistivity.

Another limitation of MHD (and fluid theories in general) is that they depend on the assumption that the plasma is strongly collisional (this is the first criterion listed above), so that the time scale of collisions is shorter than the other characteristic times in the system, and the particle distributions are Maxwellian. This is usually not the case in fusion, space and astrophysical plasmas. When this is not the case, or we are interested in smaller spatial scales, it may be necessary to use a kinetic model which properly accounts for the non-Maxwellian shape of the distribution function. However, because MHD is relatively simple and captures many of the important properties of plasma dynamics it is often qualitatively accurate and is almost invariably the first model tried.

Effects which are essentially kinetic and not captured by fluid models include double layers, Landau damping, a wide range of instabilities, chemical separation in space plasmas and electron runaway.

Dennis Gabor

Dennis Gabor, was a Hungarian electrical engineer and inventor, most notable for inventing holography, for which he later received the Nobel Prize in Physics.

Born	5 June 1900, Budapest, Hungary
Died	9 February 1979 (age 78), London, England
Fields	Electrical engineering
Institutions	Imperial College London, British Thomson-Houston
Alma mater	Technical University of Berlin, Technical University of Budapest
Known for	Invention of holography
Notable awards	Nobel Prize in Physics (1971), IEEE Medal of Honor (1970)

HOLOGRAPHY (1971)

Holography (from the Greek, whole + writing, drawing) is a technique that allows the light scattered from an object to be recorded and later reconstructed so that it appears as if the object is in the same position relative to the recording medium as it was when recorded. The image changes as the position and orientation of the viewing system changes in exactly the same way as if the object were still present, thus making the recorded image (hologram) appear three-dimensional.

The technique of holography can also be used to optically store, retrieve, and process information. While holography is commonly used to display static 3D pictures, it is not yet possible to generate arbitrary scenes by a holographic volumetric display.

Holography was invented in 1947 by the Hungarian, British physicist Dennis Gabor, work for which he received the Nobel Prize in Physics in 1971. Pioneering work in the field of physics by other scientists including Mieczysław Wolfke resolved technical issues which previously had prevented advancement. The discovery was an unexpected result of research into improving electron microscopes at the British Thomson-Houston Company in Rugby, England, and the company filed a patent in December 1947. The technique as originally invented is still used in electron microscopy, where it is known as electron holography, but holography as a light-optical technique did not really advance until the development of the laser in 1960.

The first practical optical holograms that recorded 3D objects were made in 1962 by Yuri Denisyuk in the Soviet Union and by Emmett Leith and Juris Upatnieks at University of Michigan, USA. Advances in photochemical processing techniques to produce high-quality display holograms were achieved by Nicholas J. Phillips.

Several types of holograms can be made. Transmission holograms, such as those produced by Leith and Upatnieks, are viewed by shining laser light through them and looking at the reconstructed image from the side of the hologram opposite the source. A later refinement, the rainbow transmission hologram, allows more convenient illumination by white light rather than by lasers. Rainbow holograms are commonly seen today on credit cards as a security feature and on product packaging. These versions of the rainbow transmission hologram are commonly formed as surface relief patterns in a plastic film, and they incorporate a reflective aluminum coating that provides the light from behind to reconstruct their imagery.

Another kind of common hologram, the reflection or Denisyuk hologram is capable of multicolour image reproduction using a white light illumination source on the same side of the hologram as the viewer.

John Bardeen

John Bardeen was an American physicist and electrical engineer, the first person to have won the Nobel Prize in Physics twice: first in 1956 with William Shockley and Walter Brattain for the invention of the transistor; and again in 1972 with Leon Neil Cooper and John Robert Schrieffer for a fundamental theory of conventional superconductivity known as the BCS theory.

Born	23 May 1908, Madison, Wisconsin, USA
Died	30 January 1991 (age 82), Boston, Massachusetts
Nationality	American
Fields	Physics
Institutions	Bell Labs, University of Minnesota, University of Illinois at Urbana-Champaign
Alma mater	University of Wisconsin-Madison, Princeton University
Doctoral advisor	Eugene Wigner
Doctoral students	Nick Holonyak, John Schrieffer
Known for	Transistor, BCS theory
Notable awards	Nobel Prize in Physics (1956), IEEE Medal of Honor (1971), Nobel Prize in Physics (1972)

BCS THEORY (1972)

BCS theory is the first microscopic theory of superconductivity, proposed by Bardeen, Cooper, and Schrieffer in 1957 since the discovery of superconductivity in 1911. It describes superconductivity as a microscopic effect caused by a condensation of pairs of electrons into a boson-like state.

The mid 1950s saw rapid progress in the understanding of superconductivity. It began in the 1948 paper, On the Problem of the Molecular Theory of Superconductivity where Fritz London proposed that the phenomenological

London equations may be consequences of the coherence of a quantum state. In 1953, Brian Pippard, motivated by penetration experiments, proposed that this would modify the London equations via a new scale parameter called the coherence length. John Bardeen then argued in the 1955 paper, Theory of the Meissner Effect in Superconductors that such a modification naturally occurs in a theory with an energy gap. The key ingredient was Leon Neil Cooper's calculation of the bound states of electrons subject to an attractive force in his 1956 paper, Bound Electron Pairs in a Degenerate Fermi Gas. In 1957 Bardeen and Cooper assembled these ingredients and constructed a theory, the BCS theory, with Robert Schrieffer. The theory was first announced in February 1957 in the letter, Microscopic Theory of Superconductivity. The demonstration that the phase transition is second order, that it reproduces the Meissner effect and the calculations of specific heats and penetration depths appeared in the July 1957 article, Theory of superconductivity. They received the Nobel Prize in Physics in 1972 for this theory. The 1950 Landau-Ginzburg theory of superconductivity is not cited in either of the BCS papers.

In 1986, high temperature superconductivity was discovered (i.e. superconductivity at temperatures considerably above the previous limit of about 30 K; up to about 130 K). It is believed that at these temperatures other effects are at play; these effects are not yet fully understood. It is possible that these unknown effects also control superconductivity even at low temperatures for some materials.

BCS theory starts from the assumption that there is some attraction between electrons, which can overcome the Coulomb repulsion. In most materials (in low temperature superconductors), this attraction is brought about indirectly by the coupling of electrons to the crystal lattice.

Leon N Cooper

Leon N Cooper is an American physicist and Nobel Prize laureate, who with John Bardeen and John Robert Schrieffer, developed the BCS theory of superconductivity. He is also the namesake of the Cooper pair.

Born	28 February 1930 (age 80), New York City, U.S.
Residence	United States
Nationality	United States
Fields	Physics
Institutions	Brown University
Alma mater	Columbia University
Doctoral advisor	Robert Serber
Doctoral students	Elie Bienenstock, Paul Munro, Nathan Intrator, Omer Artun, Michael Perrone Alan Saul
Known for	Superconductivity, Cooper pairs
Notable awards	Nobel Prize in Physics (1972)

COOPER PAIR (1972)

In condensed matter physics, a Cooper pair is the name given to two electrons (or other fermions) that are bound together at low temperatures in a certain manner first described in 1956 by American physicist Leon Cooper. Cooper showed that an arbitrarily small attraction between electrons in a metal can cause a paired state of electrons to have a lower energy than the Fermi energy, which implies that the pair is bound. In low temperature superconductors, this attraction is due to the electron-phonon interaction. The Cooper pair state is responsible for superconductivity, as described in the BCS theory developed by John Bardeen, John Schrieffer and Leon Cooper for which they shared the 1972 Nobel Prize.

Although Cooper pairing is a quantum effect, the reason for the pairing can be seen from a simplified classical explanation. An electron in a metal normally behaves as a free particle. The electron is repelled from other electrons due to their negative charge, but it also attracts the positive ions that make up the rigid lattice of the metal. This attraction distorts the ion lattice, moving the ions slightly toward the electron, increasing the positive charge density of the lattice in the vicinity. This positive charge can attract other electrons. At long distances this attraction between electrons due to the displaced ions can overcome the electrons' repulsion due to their negative charge, and cause them to pair up. The rigorous quantum mechanical explanation shows that the effect is due to electron-phonon interactions.

The energy of the pairing interaction is quite weak, of the order of 10^{-3} eV, and thermal energy can easily break the pairs up. So only at low temperatures a significant number of the electrons in a metal in Cooper pairs. The electrons in a pair are not necessarily close together; because the interaction is long range, paired electrons may still be many hundreds of nanometers apart. This distance is usually greater than the average interelectron distance, so many Cooper pairs can occupy the same space. Electrons have spin-$\frac{1}{2}$, so they are fermions, but a Cooper pair is a composite boson as its total spin is integer (0 or 1). This means the wave functions are symmetric under particle interchange, and they are allowed to be in the same state. The tendency for all the Cooper pairs in a body to 'condense' into the same ground quantum state is responsible for the peculiar properties of superconductivity.

The BCS theory is also applicable to other fermion systems, such as helium-3. Thus, similar pairs formed by two helium-3 atoms are also called Cooper pairs. Recently it has been shown that Cooper pairs can also be composed by two bosons. Here the pairing is supported by entanglement in an optical lattice.

John Robert Schrieffer

 John Robert Schrieffer is an American physicist and, with John Bardeen and Leon N Cooper, recipient of the 1972 Nobel Prize for Physics for developing the BCS theory, the first successful microscopic theory of superconductivity.

Born	31 May 1931 (age 80)
Nationality	United States
Fields	Physics
Institutions	University of California, Santa Barbara, University of Florida Florida State University
Alma mater	Massachusetts Institute of Technology, University of Illinois at Urbana-Champaign
Known for	Superconductivity
Notable awards	Nobel Prize in Physics (1972)

SUPERCONDUCTIVITY (1972)

BCS derived several important theoretical predictions that are independent of the details of the interaction, since the quantitative predictions mentioned below hold for any sufficiently weak attraction between the electrons and this last condition is fulfilled for many low temperature superconductors, the so-called weak-coupling case. These have been confirmed in numerous experiments:

- The electrons are bound into Cooper pairs, and these pairs are correlated due to the Pauli Exclusion Principle for the electrons, from which they are constructed. Therefore, in order to break a pair, one has to change energies of all other pairs. This means there is an energy gap for single-particle excitation, unlike in the normal metal (where the state of an electron can be changed by adding an arbitrarily small amount of energy). This energy gap is highest at low temperatures but vanishes at the transition temperature when superconductivity ceases to exist. The BCS theory gives an expression that shows how the gap grows with the strength of the attractive interaction and the (normal phase) single particle density of states at the Fermi energy. Furthermore, it describes how the density of states is changed on entering the superconducting state, where there are no electronic states any more at the Fermi energy. The energy gap is most directly observed in tunneling experiments and in reflection of microwaves from the superconductor.

- BCS theory predicts the dependence of the value of the energy gap E at temperature T on the critical temperature T_c. The ratio between the value of the energy gap at zero temperature and the value of the superconducting transition temperature (expressed in energy units) takes the universal value of 3.5, independent of material. Near the critical temperature the relation asymptotes to

$$E = 3.52 k_B T_c \sqrt{1 - (T/T_c)}$$

which is of the form suggested by M. J. Buckingham in very high frequency absorption in superconductors based on the fact that the superconducting phase transition is second order, that the superconducting phase has a mass gap and on Blevins, Gordy and Fairbank's experimental results the previous year on the absorption of millimeter waves by superconducting tin.

- Due to the energy gap, the specific heat of the superconductor is suppressed strongly at low temperatures, there being no thermal excitations left. However, before reaching the transition temperature, the specific heat of the superconductor becomes, even higher than that of the normal conductor (measured immediately above the transition) and the ratio of these two values is found to be universally given by 2.5.

Reona Esaki

 Reona Esaki also known as Leo Esaki is a Japanese physicist who shared the Nobel Prize in Physics in 1973 with Ivar Giaever and Brian David Josephson for his discovery of the phenomenon of electron tunneling.

He is known for his invention of the Esaki diode, which exploited that phenomenon. This research was done when he was with Tokyo Tsushin Kogyo (now known as Sony). He has also contributed as a pioneer of the semiconductor superlattice while he was with IBM.

Born	12 March 1925 (age 85), Osaka, Japan
Nationality	Japan
Fields	Applied physics
Known for	Electron tunneling, Esaki diode
Notable awards	Stuart Ballantine Medal (1961), Nobel Prize in Physics (1973) IEEE Medal of Honor

QUANTUM TUNNELING (1973)

Quantum tunneling refers to the phenomena of a particle's ability to penetrate energy barriers within electronic structures.

The scientific terms for this are Wave-mechanical tunneling, Quantum-mechanical tunneling and the Tunnel effect. The Tunnel Effect is an evanescent wave coupling effect that occurs in the context of quantum mechanics. Particles behave in a manner calculated with Schrödinger's wave-equations. All waves die away, but according to the laws of physics, the energy in these waves pass on. Wave coupling effects, mathematically equivalent to quantum tunneling mechanics, can occur with Maxwell's wave-equation (both with light and with microwaves), and with the common non-dispersive wave-equation often applied to waves on strings and to acoustics.

For these effects to occur there must be a situation where a thin region of medium type 2 is sandwiched between two regions of medium type 1, and the properties of these media have to be such that the wave equation has traveling wave solutions in medium type 1, but real exponential solutions (rising and falling) in medium type 2. In optics, medium type 1 might be glass, medium type 2 might be a vacuum. In quantum mechanics, in connection with motion of a particle, medium type 1 is a region of space where the particle's total energy is greater than its potential energy, medium type 2 is a region of space (known as the barrier) where the particle's total energy is less than its potential energy.

If conditions are right, amplitude from a traveling wave, incident on medium type 2 from medium type 1, can leak through medium type 2 and emerge as a traveling wave in the second region of medium type 1 on the far side. If the second region of medium type 1 is not present, then the traveling wave incident on medium type 2 is totally reflected, although it does penetrate into medium type 2 to some extent. Depending on the wave equation being used, the leaked amplitude is interpreted physically as traveling energy or as a traveling particle, and, numerically, the ratio of the square of the leaked amplitude to the square of the incident amplitude gives the proportion of incident energy transmitted out the far side, or the probability that the particle tunnels through the barrier.

Ivar Giaever

Ivar Giaever is a physicist who shared the Nobel Prize in Physics in 1973 with Leo Esaki and Brian Josephson for their discoveries regarding tunneling phenomena in solids. Giaever's share of the prize was specifically for his experimental discoveries regarding tunneling phenomena in semiconductors and superconductors.

Giaever is an institute professor emeritus at the Rensselaer Polytechnic Institute, a professor-at-large at the University of Oslo, and the president of Applied Biophysics. Giaever's experimental demonstration of tunnelling in superconductors stimulated the theoretical physicist Brian Josephson to work on the phenomenon, leading to his prediction of the Josephson Effect in 1962.

Born	5 April 1929 (age 81), Bergen, Norway
Nationality	Norway
Fields	Physics
Known for	Solid-state physics
Notable awards	Nobel Prize in Physics (1973)

SOLID-STATE PHYSICS (1973)

Solid-state physics, the largest branch of condensed matter physics, is the study of rigid matter, or solids, through methods such as quantum mechanics, crystallography, electromagnetism and metallurgy. Solid-state physics considers how the large-scale properties of solid materials result from their atomic-scale properties. Solid-state physics thus forms the theoretical basis of materials science, as well as having direct applications, for example in the technology of transistors and semiconductors.

Solid materials are formed from densely packed atoms, with intense interaction forces between them. These interactions are responsible for the mechanical, thermal, electrical, magnetic and optical properties of solids. Depending on the material involved and the conditions in which it was formed, the atoms may be arranged in a regular, geometric pattern or irregularly.

The bulk of solid-state physics theory and research is focused on crystals, largely because the periodicity of atoms in a crystal, its defining characteristic, facilitates mathematical modeling, and also because crystalline materials often have electrical, magnetic, optical, or mechanical properties that can be exploited for engineering purposes.

The forces between the atoms in a crystal can take a variety of forms. For example, in a crystal of sodium chloride, the crystal is made up of ionic sodium and chlorine, and held together with ionic bonds. In others, the atoms share electrons and form covalent bonds. In metals, electrons are shared amongst the whole crystal in metallic bonding. Finally, the noble gases do not undergo any of these types of bonding. In solid form, the noble gases are held together with van der Waals forces resulting from the polarization of the electronic charge cloud on each atom. The differences between the types of solid result from the differences between their bonding.

Many properties of materials are affected by their crystal structure. This structure can be investigated using a range of crystallographic techniques, including X-ray crystallography, neutron diffraction and electron diffraction.

The sizes of the individual crystals in a crystalline solid material vary depending on the material involved and the conditions when it was formed. Most crystalline materials encountered in everyday life are polycrystalline, with the individual crystals being microscopic in scale, but macroscopic single crystals can be produced either naturally or artificially.

Real crystals feature defects or irregularities in the ideal arrangements, and it is these defects that critically determine many of the electrical and mechanical properties of real materials.

Brian David Josephson

 Brian David Josephson, is a Welsh physicist. He became a Nobel Prize laureate in 1973 for the prediction of the eponymous Josephson Effect. He is also a fellow of Trinity College, Cambridge.

Born	4 January 1940 (age 70), Cardiff, Wales
Nationality	United Kingdom
Fields	Physics
Institutions	University of Cambridge, Trinity College, Cambridge
Known for	His work in condensed matter physics, Josephson effect
Notable awards	Nobel Prize for Physics (1973)

JOSEPHSON EFFECT (1973)

The Josephson effect is the phenomenon of current flow across two weakly coupled superconductors, separated by a very thin insulating barrier. This arrangement, two superconductors linked by a non-conducting barrier is known as a Josephson junction; the current that crosses the barrier is the Josephson current. The terms are named after British physicist Brian David Josephson, who predicted the existence of the effect in 1962. It has important applications in quantum-mechanical circuits, such as SQUIDs or RSFQ digital electronics.

The Josephson junction can be made in various physical configurations, for instance as Dayem bridge.

The basic equations governing the dynamics of the Josephson effect are:

$U(t) = \dfrac{\hbar}{2e} \dfrac{\partial \phi}{\partial t}$ (superconducting phase evolution equation)

$I(t) = I_c \sin(\phi(t))$ (Josephson or weak-link current-phase relation).

where $u(t)$ and $I(t)$ are the voltage and current across the Josephson junction, $\phi(t)$ is the phase difference across the junction (i.e., the difference in phase factor, or equivalently, argument, between the Ginzburg-Landau complex order parameter of the two superconductors composing the junction), and I_c is a constant, the *critical current* of the junction. The critical current is an important phenomenological parameter of the device that can be affected by temperature as well as by an applied magnetic field. The physical constant, $h/2e$ is the magnetic flux quantum, the inverse of which is the Josephson constant.

The three main effects predicted by Josephson follow from these relations:

The DC Josephson effect: This refers to the phenomenon of a direct current crossing from the insulator in the absence of any external electromagnetic field, owing to tunneling. This DC Josephson current is proportional to the sine of the phase difference across the insulator, and may take values between $-I_c$ and I_c.

The AC Josephson effect. With a fixed voltage U_{DC} across the junctions, the phase will vary linearly with time and the current will be an AC current with amplitude I_c and frequency $\dfrac{2e}{h}.U_{DC}$. The complete expression for the current drive I_{ext} becomes $I_{ext} = C_j \dfrac{dv}{dt} + I_j \sin\phi + \dfrac{V}{R}$ This means a Josephson junction can act as a perfect voltage-to-frequency converter.

The inverse AC Josephson effect: If the phase takes the form

$\phi(t) = \phi_0 + nwt + a\sin(wt)$, the voltage and current will be

$$U(t) = \frac{\hbar}{2e} w(n + a\cos(wt)), \quad I(t) = I_c \sum_{m=-\infty}^{\infty} J_n(a)\sin(\phi_0 + (n+m)wt)$$

The DC components will then be

$$U_{DC} = n\frac{\hbar}{2e} w, \quad I(t) = I_c J_{-n}(a)\sin\phi_0$$

hence, for distinct DC voltages, the junction may carry a DC current and the junction acts like a perfect frequency-to-voltage converter.

Sir Martin Ryle

Sir Martin Ryle was an English radio astronomer who developed revolutionary radio telescope systems and used them for accurate location and imaging of weak radio sources.

In 1946 Ryle and Vonberg were the first people to publish interferometric astronomical measurements at radio wavelengths, although it is claimed that Joseph Pawsey from the University of Sydney had actually made interferometric measurements earlier in the same year. With improved equipment, Ryle observed the most distant known galaxies in the universe at that time. He was the first Professor of Radio Astronomy at the University of Cambridge, and founding director of the Mullard Radio Astronomy Observatory. He was Astronomer Royal from 1972 to 1982. Ryle and Antony Hewish shared the Nobel Prize for Physics in 1974, the first Nobel Prize awarded in recognition of astronomical research.

Born	27 September 1918, Brighton, England
Died	14 October 1984 (age 66), Cambridge, England
Nationality	United Kingdom
Fields	Astronomy
Doctoral advisor	J. A. Ratcliffe
Known for	Radio astronomy
Notable awards	Hughes Medal (1954), RAS Gold Medal (1964), Henry Draper Medal (1965), Royal Medal (1973), Bruce Medal (1974), Nobel Prize in Physics (1974)

RADIO ASTRONOMY (1974)

Radio astronomy is a subfield of astronomy that studies celestial objects at radio frequencies. The initial detection of radio waves from an astronomical object was made in the 1930s, when Karl Jansky observed radiation coming from the

Milky Way. Subsequent observations have identified a number of different sources of radio emission. These include stars and galaxies, as well as entirely new classes of objects, such as radio galaxies, quasars, pulsars, and masers. The discovery of the cosmic microwave background radiation, which provided compelling evidence for the Big Bang, was made through radio astronomy.

Radio astronomy is conducted using large radio antennae referred to as radio telescopes, that are either used singularly, or with multiple linked telescopes utilizing the techniques of radio interferometry and aperture synthesis. The use of interferometry allows radio astronomy to achieve high angular resolution, as the resolving power of an interferometer is set by the distance between its components, rather than the size of its components.

Before Jansky observed the Milky Way in the 1930s, physicists speculated that radio waves could be observed from astronomical sources. In the 1860s, James Clerk Maxwell's equations had shown that electromagnetic radiation is associated with electricity and magnetism, and could exist at any wavelength. Several attempts were made to detect radio emission from the Sun by experimenters such as Nikola Tesla and Oliver Lodge, but those attempts were unable to detect any emission due to technical limitations of their instruments.

Karl Jansky made the discovery of the first astronomical radio source serendipitously in the early 1930s. As an engineer with Bell Telephone Laboratories, he was investigating static that interfered with short wave transatlantic voice transmissions. Using a large directional antenna, Jansky noticed that his analog pen-and-paper recording system kept recording a repeating signal of unknown origin. Since the signal peaked about every 24 hours, Jansky originally suspected the source of the interference was the Sun crossing the view of his directional antenna.

Antony Hewish

Antony Hewish is a British radio astronomer who won the Nobel Prize for Physics in 1974 (together with fellow radio astronomer Martin Ryle) for his work on the development of radio aperture synthesis and its role in the discovery of pulsars. (Jocelyn Bell Burnell, Hewish's graduate student, was not recognized, although she was the first to notice the stellar radio source that was later recognised as a pulsar.) He was also awarded the Eddington Medal of the Royal Astronomical Society in 1969.

Born	11 May 1924 (age 86) Fowey, Cornwall
Nationality	United Kingdom
Fields	Radio astronomy
Known for	Pulsars
Notable awards	Nobel Prize in Physics (1974), Eddington Medal of the Royal Astronomical Society (1969)

PULSAR (1974)

Pulsars are highly magnetized, rotating neutron stars that emit a beam of electromagnetic radiation. The radiation can only be observed when the beam of emission is pointing towards the Earth. This is called the lighthouse effect and gives rise to the pulsed nature that gives pulsars their name. Because neutron stars are very dense objects, the rotation period and thus the interval between observed pulses is very regular. For some pulsars, the regularity of pulsation is as precise as an atomic clock. The observed periods of their pulses range from 1.4 milliseconds to 8.5 seconds. A few pulsars are known to have planets orbiting them, such as PSR B1257+12. Werner Becker of the Max Planck Institute for Extraterrestrial Physics said in 2006. The theory of how pulsars emit their radiation is still in its infancy, even after nearly forty years of work.

The first pulsar was observed on 28 November, 1967 by Jocelyn Bell Burnell and Antony Hewish. Initially baffled as to the seemingly unnatural regularity of its emissions, they dubbed their discovery LGM-1, for little green men (a name for intelligent beings of extraterrestrial origin). While the hypothesis that pulsars were beacons from extraterrestrial civilizations was never taken very seriously, some discussed the far reaching implications if it turned out to be true. Their pulsar was later dubbed CP 1919, and is now known by a number of designators including PSR 1919+21, PSR B1919+21 and PSR J1921 + 2153. Although CP 1919 emits in radio wavelengths, pulsars have, subsequently, been found to emit in visible light, X-ray, and/or gamma ray wavelengths.

The word pulsar is a contraction of pulsating star, and first appeared in print in 1968.

An entirely novel kind of star came to light and was referred to, by astronomers, as LGM (Little Green Men). Now it is thought to be a novel type between a white dwarf and a neutron. The name Pulsar is likely to be given to it. Dr. A. Hewish told me yesterday: ... I am sure that today every radio telescope is looking at the Pulsars.

The suggestion that pulsars were rotating neutron stars was put forth independently by Thomas Gold and Franco Pacini in 1968, and was soon proven beyond reasonable doubt by the discovery of a pulsar with a very short (33-millisecond) pulse period in the Crab nebula.

In 1974, Antony Hewish became the first astronomer to be awarded the Nobel Prize in physics.

Aage Niels Bohr

Aage Niels Bohr was a Danish nuclear physicist and Nobel laureate, and the son of Niels and Margrethe Bohr. Bohr, Mottelson and Rainwater were jointly awarded the 1975 Nobel Prize in Physics for the discovery of the connection between collective motion and particle motion in atomic nuclei and the development of the theory of the structure of the atomic nucleus based on this connection.

Born	19 June 1922, Copenhagen, Denmark
Died	8 September 2009 (age 87), Copenhagen, Denmark
Nationality	Danish
Fields	Nuclear physicist
Institutions	Manhattan Project, University of Copenhagen
Alma mater	University of Copenhagen
Known for	Geometry of atomic nuclei
Notable awards	Wetherill Medal (1974), Nobel Prize in Physics (1975)
Aage Bohr is the son of noted physicist Niels Bohr	

ATOMIC NUCLEUS

The nucleus is the very dense region consisting of nucleons (protons and neutrons) at the center of an atom. Almost all of the mass in an atom is made up from the protons and neutrons in the nucleus, with a very small contribution from the orbiting electrons. It was discovered in 1911, as a result of Ernest Rutherford's interpretation of the famous 1909 Rutherford experiment performed by Hans Geiger and Ernest Marsden, under the direction of Rutherford.

The diameter of the nucleus is in the range of 1.6 fm (1.6 × 10^{-15} m) for hydrogen (the diameter of a single proton) to about 15 fm for the heaviest atoms, such as uranium. These dimensions are much smaller than the diameter of the atom itself (nucleus + electronic cloud), by a factor of about 23,000 (uranium) to about 145,000 (hydrogen).

The branch of physics concerned with studying and understanding the atomic nucleus, including its composition and the forces which bind it together, is called nuclear physics.

The term nucleus is from Latin *nucleus* (kernel), derived from *nux* (nut). In 1844, Michael Faraday used the term to refer to the central point of an atom. The modern atomic meaning as proposed by Ernest Rutherford in 1912. The adoption of the term nucleus to atomic theory, however, was not immediate. In 1916, for example, Gilbert N. Lewis stated, in his famous article *The Atom and the Molecule,* that the atom is composed of the *kernel* and an outer atom or *shell.*

There are many different historical models of the atomic nucleus, none of which to this day completely explains experimental data on nuclear structure. A useful review of 37 known models of the atomic nucleus is provided by Cook.

The nuclear radius (R) is considered to be one of the basic things that any model must predict. For stable nuclei (not halo nuclei or other unstable distorted nuclei) the nuclear radius is roughly proportional to the cube root of the mass number (A) of the nucleus, and particularly in nuclei containing many nucleons, as they arrange in more spherical configurations.

Ben Roy Mottelson

Ben Roy Mottelson is a Danish American nuclear physicist. He won the 1975 Nobel Prize in Physics for his work on the non-spherical geometry of atomic nuclei. Rainwater, Bohr and Mottelson were jointly awarded the 1975 Nobel Prize in Physics for the discovery of the connection between collective motion and particle motion in atomic nuclei and the development of the theory of the structure of the atomic nucleus based on this connection.

Born	9 July 1926, Chicago, Illinois
Nationality	Danish and American
Fields	Nuclear physicist
Institutions	Nordita
Alma mater	Purdue University, Harvard University
Doctoral advisor	Julian Schwinger
Known for	Geometry of atomic nuclei
Notable awards	Wetherill Medal (1974), Nobel Prize in Physics (1975)

ATOMIC NUCLEUS (1975)

The nucleus of an atom consists of protons and neutrons (two types of baryons) bound by the nuclear force (also known as the residual strong force). These baryons are further composed of subatomic fundamental particles known as quarks bound by the strong interaction. Which chemical element an atom represents is determined by the number of protons in the nucleus. Each proton carries a single positive charge, and the total electrical charge of the nucleus is spread fairly uniformly throughout its body, with a fall-off at the edge.

Major exceptions to this rule are the light elements hydrogen and helium, where the charge is concentrated most highly at the single central point (without a volume of uniform charge), as would be expected for fermions (in this case, protons) in 1s states without orbital angular momentum.

As each proton carries a unit of charge, the charge distribution is indicative of the proton distribution. The neutron distribution probably is similar.

Protons and neutrons are fermions, with different values of the isospin quantum number, so two protons and two neutrons can share the same space wave function since they are not identical quantum entities. They sometimes are viewed as two different quantum states of the same particle, the *nucleon*. Two fermions, such as two protons, or two neutrons, or a proton + neutron (the deuteron) can exhibit bosonic behavior when they become loosely bound in pairs.

In the rare case of a hyper nucleus, a third baryon called a hyperon, with a different value of the strangeness quantum number can also share the wave function. However, the latter type of nuclei are extremely unstable and are not found on Earth except in high energy physics experiments.

The neutron has a positively charged core of radius ≈ 0.3 fm surrounded by a compensating negative charge of radius between 0.3 fm and 2 fm. The proton has an approximately exponentially decaying positive charge distribution with a mean square radius of about 0.8 fm.

The stable nucleus has approximately a constant density and therefore the nuclear radius R can be approximated by the following formula, $R = r_0 A^{1/3}$

where A = Atomic mass number (the number of protons, Z, plus the number of neutrons, N) and $r_0 = 1.25$ fm $= 1.25 \times 10^{-15}$ m. In this equation, the constant r_0 varies by 0.2 fm, depending on the nucleus in question, but this is less than 20% change from a constant.

In other words, packing protons and neutrons in the nucleus gives *approximately* the same total size result as packing hard spheres of a constant size (like marbles) into a tight spherical or semi-spherical bag (some stable nuclei are not quite spherical, but are known to be prolate).

Leo James Rainwater

Leo James Rainwater was an American physicist who shared the Nobel Prize for Physics in 1975, for his part in determining the asymmetrical shapes of certain atomic nuclei.

Born	9 December 1917, Council, Idaho
Died	31 May 1986 (age 68)
Institutions	Columbia University, Manhattan Project
Alma mater	Columbia University, Caltech
Notable awards	Nobel Prize in Physics (1975), Ernest Orlando Lawrence Award (1963)

ATOMIC NUCLEUS (1975)

Nuclei are bound together by the residual strong force (nuclear force). The residual strong force is minor residum of the strong interaction which binds quarks together to form protons and neutrons. This force is much weaker *between* neutrons and protons because it is mostly neutralized within them, in the same way that electromagnetic forces *between* neutral atoms are much weaker than the electromagnetic forces that hold the parts of the atoms internally together.

The nuclear force is highly attractive at very small distances, and this overwhelms the repulsion between protons which is due to the electromagnetic force, thus allowing nuclei to exist. However, because the residual strong force has a limited range because it decays quickly with distance, only nuclei smaller than a certain size can be completely stable. The largest known completely stable nucleus is lead-208 which contains a total of 208 nucleons (126 neutrons and 82 protons). Nuclei larger than this maximal size of 208 particles are unstable and become increasingly short-lived with larger size, as the number of neutrons and protons which compose them increases beyond this number. However, bismuth-209 is also stable to beta decay and has the longest half-live to alpha decay of any known isotope, estimated at longer than the age of the universe.

The residual strong force is effective over a very short range and causes an attraction between any pair of nucleons. It is also effective for the stability of one 3-body nucleon system [PNP], helium-3, while the triton [NPN] is unstable and decays to helium-3.

The effective absolute limit of the range of the strong force is represented by halo nuclei such as lithium-11 or boron-14, in which dineutrons, or other collections of neutrons, orbit at distances of about ten fermis. These nuclei are not maximally dense. Halo nuclei form at the extreme edges of the chart of the nuclides, the neutron drip line and proton drip line, and are all unstable with short half-lives, measured in milliseconds; for example, lithium-11 has a half-life of less than 8.6 milliseconds.

Halos in effect represent an excited state with nucleons in an outer quantum shell which has unfilled energy levels below it. The halo may be made of either neutrons [NN, NNN] or protons [PP, PPP]. Examples: Nuclei which have a single neutron halo include 11Be and 19C. A two-neutron halo is exhibited by ^{6}He, ^{11}Li, ^{17}B, ^{19}B and ^{22}C. Two-neutron halo nuclei break into three fragments, never two, and are called *Borromean* because of this behavior. ^{8}He and ^{14}Be both exhibit a four-neutron halo. Nuclei which have a proton halo include ^{8}B and ^{26}P. A two-proton halo is exhibited by ^{17}Ne and ^{27}S. Proton halos are expected to be more rare and unstable than the neutron examples, because of the repulsive electromagnetic forces of the excess proton(s).

Burton Richter

 Burton Richter is a Nobel Prize winning American physicist. He led the Stanford Linear Accelerator Center (SLAC) team which co-discovered the J/ψ meson in 1974, alongside the Brookhaven National Laboratory (BNL) team lead by Samuel Ting.

This discovery was part of the so called November Revolution of particle physics. He was the SLAC director from 1984 to 1999. This discovery was also made by the team lead by Samuel Ting at Brookhaven National Laboratory, but he called the particle J. The particle thus became known as the J/ψ meson. Burton and Ting were jointly awarded the 1976 Nobel Prize in Physics for their work.

Born	22 March 1931 (age 79), Brooklyn, New York City
Nationality	American
Institutions	Stanford University, Stanford Linear Accelerator Center
Alma mater	MIT
Known for	J/ψ meson
Notable awards	Nobel Prize in Physics (1976)

J/Ψ MESON (1976)

The background to the discovery of the J/ψ was both theoretical and experimental. In the sixties, the first quark models of elementary particle physics were proposed, which said that protons, neutrons and all other baryons, and also all mesons, are made from three kinds of fractionally charged particles, the quarks, that come in three different types or flavors, called up, down, and strange. Despite the impressive ability of quark models to bring order to the elementary particle zoo, their status was considered something like mathematical fiction at the time, a simple artifact of deeper physical reasons.

Starting in 1969, deep inelastic scattering experiments at SLAC revealed surprising experimental evidence for particles inside protons. Whether these were quarks or something else was not known at first. Many experiments were needed to fully identify the properties of the subprotonic components. To a first approximation, they were indeed the already described quarks.

On the theoretical front, gauge theories with broken symmetry became the first fully viable contenders for explaining the weak interaction, after Gerardus 't Hooft discovered in 1971 how to calculate with them beyond tree level. The first experimental evidence for these electroweak unification theories was the discovery of the weak neutral current in 1973. Gauge theories with quarks became a viable contender for the strong interaction in 1973 when the concept of asymptotic freedom was identified.

However, a naive mixture of electroweak theory and the quark model led to calculations about known decay modes that contradicted observation: in particular, it predicted Z boson-mediated flavor-changing decays of a strange quark into a down quark, which were not observed. A 1970 idea of Sheldon Glashow, John Iliopoulos, and Luciano Maiani, known as the GIM mechanism, showed that the flavor-changing decays would be eliminated if there were a fourth quark, charm, that paired with the strange quark. This work led, by the summer of 1974, to theoretical predictions of what a charm/anticharm meson would be like. These predictions were ignored. The work of Richter and Ting was done for other reasons, mostly to explore new energy regimes.

The J/ψ is a subatomic particle, a flavor-neutral meson consisting of a charm quark and a charm antiquark. Mesons formed by a bound state of a charm quark and a charm anti-quark are generally known as charmonium. The J/ψ is the first excited state of charmonium (i.e, the form of the charmonium with the second-smallest rest mass). The J/ψ has a rest mass of 3,096.9 Mev/c^2, and a mean lifetime of 7.2×10^{-21} s. This lifetime was about a thousand times longer than expected.

Samuel Chao Chung Ting

Samuel Chao Chung Ting is an American physicist who received the Nobel Prize in 1976, with Burton Richter, for discovering the subatomic J/ψ particle. He is the principal investigator for the international Alpha Magnetic Spectrometer project scheduled for installation on the International Space Station in 2010.

They were chosen for the award, in the words of the Nobel committee, for their pioneering work in the discovery of a heavy elementary particle of a new kind. The discovery was made in 1974 when Ting was heading a research team at the Brookhaven National Laboratory exploring new regimes of high energy particle physics.

Born	27 January 1936 (age 74), Ann Arbor, Michigan, USA
Nationality	United States
Fields	Physics
Institutions	CERN, Columbia University, MIT
Alma mater	University of Michigan
Doctoral advisor	L.W. Jones, M.L. Perl
Known for	Discovery of the J/ψ particle
Notable awards	Nobel Prize in Physics (1976), Ernest Orlando Lawrence Award (1975) De Gasperi Award (1988)

J/Ψ MESON (1976)

Because of the nearly simultaneous discovery, the J/ψ is the only elementary particle to have a two-letter name. Richter named it SP, after the SPEAR accelerator used at SLAC; however, none of his coworkers liked that name. After consulting with Greek born Leo Resvanis to see which Greek letters were still available, and rejecting iota because its name implies insignificance, Richter chose psi - a name which, as Gerson

Goldhaber pointed out, contains the original name SP, but in reverse order. Coincidentally, later spark chamber pictures often resembled the psi shape. Ting assigned the name J to it, which is one letter removed from K, the name of the already-known strange meson; possibly by coincidence, J strongly resembles the Chinese character for Ting's name (丁).

Since the scientific community considered it unjust to give one of the two discoverers priority, most subsequent publications have referred to the particle as the J/ψ.

The first excited state of the J/ψ was called the ψ'. It is now termed the ψ (2S) or occasionally ψ(3686), indicating respectively its quantum state or mass in Mev. Other vector charm-anticharm states are denoted similarly with ψ and the quantum state (if known) or the mass. The J is not used, since Richter's group alone first found excited states.

The name charmonium is used for the J/ψ and other charm-anticharm bound states. This is by analogy with positronium, which also consists of a particle and its antiparticle.

In a hot QCD medium, when the temperature is raised well beyond the Hagedorn temperature, the J/ψ and its excitations are expected to melt. This is one of the predicted signals of the formation of the quark-gluon plasma. Heavy-ion experiments at CERN's Super Proton Synchrotron and at BNL's Relativistic Heavy Ion Collider have studied this phenomenon without a conclusive outcome as of 2009. This is due to the requirement that the disappearance of J/ψ mesons is evaluated with respect to the baseline provided by the total production of all charm quark-containing subatomic particles, and because it is widely expected that some of the J/ψ are produced and/or destroyed at time of QGP hadronization. Thus there is uncertainty in the prevailing conditions at the initial collisions.

Philip Warren Anderson

 Philip Warren Anderson is an American physicist and Nobel laureate. Anderson has made contributions to the theories of localization, antiferromagnetism and high-temperature superconductivity.

In 1977 Anderson was awarded the Nobel Prize in Physics for his investigations into the electronic structure of magnetic and disordered systems, which allowed for the development of electronic switching and memory devices in computers. Co-researchers Sir Nevill Francis Mott and John van Vleck shared the award with him. In 1982, he was awarded the National Medal of Science. He retired from Bell Labs in 1984 and is currently Joseph Henry Professor of Physics at Princeton University.

Born	13 December 1923 (age 86), Indianapolis, Indiana, USA
Nationality	United States
Fields	Physics
Institutions	Bell Laboratories, Princeton University, Cambridge University
Alma mater	Harvard University, U.S. Naval Research Laboratory
Doctoral advisor	John Hasbrouck van Vleck
Notable awards	Nobel Prize in Physics (1977)

ANDERSON LOCALIZATION (1977)

In stochastic processes, Anderson localization, also known as strong localization, is the absence of diffusion of waves in a disordered medium. This phenomenon is named after the American physicist P. W. Anderson, who is the first one to suggest the possibility of electron localization inside a semiconductor, provided that the degree of randomness of the impurities or defects is sufficiently large. Anderson localization is a general wave phenomenon that applies to the transport of electromagnetic waves, acoustic waves, quantum waves, spin waves, etc. This phenomenon is to be distinguished from weak localization, which is the precursor effect of Anderson localization. This phenomenon finds its origin in the wave interference between multiple-scattering paths. In the strong scattering limit, the severe interferences can completely halt the waves inside the disordered medium. Localized states have been predicted but never observed to easily exist inside bandgaps upon structural disorders in periodic structures.

For non-interacting electrons, a highly successful approach was put forward in 1979 by Abrahams *et al.* This scaling hypothesis of localization suggests that a metal-insulator transition (MIT) exists for non-interacting electrons in three-dimensions (3D) at zero magnetic field and in the absence of spin-orbit coupling. Much further work has subsequently supported these scaling arguments both analytically and numerically. In 1D and 2D, the same hypothesis shows that there are no extended states and thus no MIT. However, since 2 is the lower critical dimension of the localization problem, the 2D case is in a sense close to 3D: states are only marginally localized for weak disorder and a small magnetic field or spin-orbit coupling can lead to the existence of extended states and thus an MIT. Consequently, the localization lengths of a 2D system with potential disorder can be quite large so that in numerical approaches one can always find a localization-delocalization transition when decreasing either system size for fixed disorder or disorder for fixed system size.

Most numerical approaches to the localization problem use the standard tight-binding Anderson Hamiltonian with onsite potential disorder. Characteristics of the electronic eigenstates are then investigated by studies of participation numbers obtained by exact diagonalization, multifractal properties, level statistics and many others. Especially fruitful is the transfer-matrix method (TMM) which allows a direct computation of the localization lengths and further validates the scaling hypothesis by a numerical proof of the existence of a one-parameter scaling function.

Sir Nevill Francis Mott

Sir Nevill Francis Mott, was an English physicist. He won the Nobel Prize for Physics in 1977 for his work on the electronic structure of magnetic and disordered systems. The award was shared with Philip W. Anderson and J. H. Van Vleck, who had pursued independent research.

Born	30 September 1905, Leeds, England
Died	8 August 1996 (age 90), Milton Keynes, Buckinghamshire, England
Nationality	United Kingdom
Fields	Physics
Institutions	University of Manchester, Gonville and Caius College, Cambridge University of Bristol
Alma mater	St John's College, Cambridge
Notable awards	Nobel Prize in Physics (1977)

AMORPHOUS SOLID (1977)

An amorphous solid is a solid in which there is no long-range order of the positions of the atoms. (Solids in which there is long-range atomic order are called crystallines or a morphous). Most classes of solid materials can be found or prepared in an amorphous form. For instance, common window glass is an amorphous solid, many polymers (such as polystyrene) are amorphous, and even foods such as cotton candy are amorphous solids.

In principle, given a sufficiently high cooling rate, any liquid can be made into an amorphous solid. Cooling reduces molecular mobility. If the cooling rate is faster than the rate at which molecules can organize into a more thermodynamically favorable crystalline state, then an amorphous solid will be formed. Because of entropy considerations, many polymers can be made amorphous solids by cooling even at slow rates.

In contrast, if molecules have sufficient time to organize into a structure with two- or three-dimensional order, then a crystalline (or semi-crystalline) solid will be formed. Water is one example. Because of its small molecular size and ability to quickly rearrange, it cannot be made amorphous without resorting to specialized hyper quenching techniques.

Amorphous materials can also be produced by additives which interfere with the ability of the primary constituent to crystallize. For example, addition of soda to silicon dioxide results in window glass, and the addition of glycols to water results in a vitrified solid.

Some materials, such as metals, are difficult to prepare in an amorphous state. Unless a material has a high melting temperature (as ceramics do) or a low crystallization energy (as polymers tend to), cooling must be done extremely rapidly. As the cooling is performed, the material changes from a super cooled liquid, with properties one would expect from a liquid state material, to a solid. The temperature at which this transition occurs is called the glass transition temperature or T_g.

It is difficult to make a distinction between truly amorphous solids and crystalline solids if the size of the crystals is very small. Even amorphous materials have some short-range order at the atomic length scale due the nature of chemical bonding. Furthermore, in very small crystals a large fraction of the atoms are located at or near the surface of the crystal; relaxation of the surface and interfacial effects distort the atomic positions, decreasing the structural order. Even the most advanced structural characterization techniques, such as X-ray diffraction and transmission electron microscopy, have difficulty in distinguishing between amorphous and crystalline structures on these length scales.

John Hasbrouck Van Vleck

 John Hasbrouck Van Vleck was an American physicist and mathematician, co-awarded the 1977 Nobel Prize in Physics, for his contributions to the understanding of the behavior of electrons in magnetic solids.

Born	13 March 1899, Middletown, Connecticut
Died	27 October 1980, Cambridge, Massachusetts
Nationality	United States
Fields	Physics
Institutions	University of Minnesota, University of Wisconsin–Madison Harvard University, University of Oxford, Balliol College
Alma mater	University of Wisconsin-Madison, Harvard
Doctoral advisor	Edwin C. Kemble
Doctoral students	Robert Serber, Edward Mills Purcell, Philip Anderson
Notable awards	Elliott Cresson Medal (1971), Nobel Prize in Physics (1977)

ANTIFERROMAGNETISM (1977)

In materials that exhibit antiferromagnetism, the magnetic moments of atoms or molecules, usually related to the spins of electrons, align in a regular pattern with neighboring spins (on different sub lattices) pointing in opposite directions. This is, like ferromagnetism and ferrimagnetisms, a manifestation of ordered magnetism. Generally, antiferromagnetic order may exist at sufficiently low temperatures, vanishing at and above a certain temperature, the Néel temperature (named after Louis Néel, who had first identified this type of magnetic ordering). Above the Néel temperature, the material is typically paramagnetic.

An antiferromagnetic interaction acts to anti-align neighboring spins. If the energy is expressed as the sum of all pairs, i, j, over an interaction term J (i, j), times the spin of atom i times the spin of atom j, J < 0 is a ferromagnetic interaction and J > 0 is an antiferromagnetic interaction. The combination of both can lead to spin glass behavior.

When no external field is applied, the antiferromagnetic structure corresponds to a vanishing total magnetization. In a field, a kind of ferrimagnetic behavior may be displayed in the antiferromagnetic phase, with the absolute value of one of the sublattice magnetizations differing from that of the other sublattice, resulting in a nonzero net magnetization.

The magnetic susceptibility of an antiferromagnetic material typically shows a maximum at the Néel temperature. In contrast, at the transition between the ferromagnetic to the paramagnetic phases the susceptibility will diverge. In the antiferromagnetic case, a divergence is observed in the staggered susceptibility.

Various microscopic (exchange) interactions between the magnetic moments or spins may lead to antiferromagnetic structures. In the simplest case, one may consider an Ising model on an bipartite lattice, e.g. the simple cubic lattice, with couplings between spins at nearest neighbor sites. Depending on the sign of that interaction, ferromagnetic or antiferromagnetic order will result. Geometrical frustration or competing ferro- and antiferromagnetic interactions may lead to different and, perhaps, more complicated magnetic structures.

Antiferromagnetic materials occur commonly among transition metal compounds, especially oxides. An example is the heavy-fermion superconductor URu_2Si_2. Better known examples include hematite, metals such as chromium, alloys such as iron manganese (FeMn), and oxides such as nickel oxide (NiO). There are also numerous examples among high nuclearity metal clusters. Organic molecules can also exhibit antiferromagnetic coupling under rare circumstances, as seen in radicals such as 5-dehydro-m-xylylene.

Pyotr Leonidovich Kapitsa

Pyotr Leonidovich Kapitsa was an innovative Soviet Russian physicist and Nobel laureate, who made important discoveries in several different areas. Belatedly (1978), Kapitsa won the Nobel Prize in Physics for the work in low temperature physics that he did about 1937. He shared this prize with Arno Allan Penzias and Robert Woodrow Wilson (who won for work unrelated to Kapitsa's).

Born	9 July 1894, Kronstadt, Russian Empire
Died	8 April 1984 (age 89)
Citizenship	Russia, USSR
Nationality	Russian
Fields	Physics
Known for	Superfluidity
Notable awards	Franklin Medal (1944), Nobel Prize in Physics (1978)

SUPER FLUID (1978)

Super fluidity is a phase of matter or description of heat capacity in which unusual effects are observed when liquids, typically of helium-4 or helium-3, overcome friction by surface interaction when at a stage (known as the λ point, which is temperature and pressure, for helium-4) at which the liquid's viscosity becomes zero. Also known as a major facet in the study of quantum hydrodynamics, it was discovered by Pyotr Kapitsa, John F. Allen, and Don Misener in 1937 and has been described through phenomenological and microscopic theories. In the 1950s Hall and Vinen performed experiments establishing the existence of quantized vortex lines. In the 1960s, Rayfield and Reif established the existence of quantized vortex rings. Packard has observed the intersection of vortex lines with the free surface of the fluid, and Avenel and Varoquaux have studied the Josephson Effect in super fluid ^{4}He.

Although the phenomenologies of the super fluid states of helium-4 and helium-3 are very similar, the microscopic details of the transitions are very different. Helium-4 atoms are bosons, and their super fluidity can be understood in terms of the Bose statistics that they obey. Specifically, the super fluidity of helium-4 can be regarded as a consequence of Bose-Einstein condensation in an interacting system. On the other hand, helium-3 atoms are fermions, and the super fluid transition in this system is described by a generalization of the BCS theory of superconductivity. In it, Cooper pairing takes place between atoms rather than electrons, and the attractive interaction between them is mediated by spin fluctuations rather than phonons. A unified description of superconductivity and super fluidity is possible in terms of gauge symmetry breaking.

Super fluids, such as super cooled helium-4, exhibit many unusual properties. Super fluid acts as if it were a mixture of a normal component, with all the properties associated with normal fluid, and a super fluid component. The super fluid component has zero viscosity, zero entropy, and infinite thermal conductivity. It is thus impossible to set up a temperature gradient in a super fluid, much as it is impossible to set up a voltage difference in a superconductor. Application of heat to a spot in super fluid helium results in a wave of heat conduction at the relatively high velocity of 20 m/s, called second sound.

Although the phenomenologies of the super fluid states of helium-4 and helium-3 are very similar, the microscopic details of the transitions are very different. Helium-4 atoms are bosons, and their super fluidity can be understood in terms of the Bose-Einstein statistics that they obey. Specifically, the super fluidity of helium-4 can be regarded as a consequence of Bose-Einstein condensation in an interacting system. On the other hand, helium-3 atoms are fermions, and the super fluid transition in this system is described by a generalization of the BCS theory of superconductivity. In it, Cooper pairing takes place between atoms rather than electrons, and the attractive interaction between them is mediated by spin fluctuations rather than phonons. A unified description of superconductivity and super fluidity is possible in terms of gauge symmetry breaking.

Arno Allan Penzias

Arno Allan Penzias is an American physicist and Nobel laureate in physics. Penzias was born in Munich, Germany. In 1946, Penzias became a naturalized citizen of the United States.

He graduated from Brooklyn Technical High School in 1951 and received a bachelor's degree from the City College of New York in 1954. From Columbia University, he received his Master's degree in 1958 and his Ph.D. in 1962. Penzias and Wilson received the 1978 Nobel Prize, sharing it with Pyotr Leonidovich Kapitsa.

Born	26 April 1933 (age 77), Munich, Germany
Nationality	United States
Fields	Physics
Institutions	Bell Labs
Alma mater	CCNY, Columbia University
Known for	Cosmic microwave background radiation
Notable awards	Nobel Prize in Physics (1978)

COSMIC MICROWAVE BACKGROUND RADIATION (1978)

In cosmology, cosmic microwave background (CMB) radiation (also CMBR, CBR, MBR, and relic radiation) is a form of electromagnetic radiation filling the universe. With a traditional optical telescope, the space between stars and galaxies (the background) is pitch black. But with a radio telescope, there is a faint background glow, almost exactly the same in all directions, that is not associated with any star, galaxy, or other object. This glow is strongest in the microwave region of the radio spectrum, hence the name cosmic microwave background radiation. The CMB's discovery in

1964 by American radio astronomers Arno Penzias and Robert Wilson was the culmination of work initiated in the 1940s, and earned them the 1978 Nobel Prize.

The CMBR is well explained as radiation left over from an early stage in the creation of the universe, and its discovery is considered a landmark confirmation of the Big Bang model of the universe. When the universe was young, before the formation of stars and planets, it was smaller, much hotter, and filled with a uniform glow from its white-hot fog of hydrogen plasma. As the universe expanded, both the plasma and the radiation filling it, grew cooler. When the universe cooled enough, stable atoms could form. These atoms could no longer absorb the thermal radiation, and the universe became transparent instead of being an opaque fog. The photons that existed at that time have been propagating ever since, though growing fainter and less energetic, since the exact same photons fill a greater and greater universe. This is the source for the term relic radiation, another name for the CMBR.

Precise measurements of cosmic background radiation are critical to cosmology, since any proposed model of the universe must explain this radiation. The CMBR has a thermal black body spectrum at a temperature of 2.725 K, thus the spectrum peaks in the microwave range frequency of 160.2 GHz, corresponding to a 1.9 mm wavelength. The glow is almost but not quite uniform in all directions, and shows a very specific pattern equal to that expected if the inherent randomness of a red-hot gas is blown up to the size of the universe. In particular, the spatial power spectrum (how much difference is observed versus how far apart the regions are on the sky) contains small anisotropies, or irregularities, which vary with the size of the region examined. They have been measured in detail, and match what would be expected if small thermal fluctuations had expanded to the size of the observable space we can detect today. This is still a very active field of study, with scientists seeking both better data (for example, the Planck spacecraft) and better interpretations of the initial conditions of expansion.

Although many different processes might produce the general form of a black body spectrum, no model other than the Big Bang has yet explained the fluctuations. As a result, most cosmologists consider the Big Bang model of the universe to be the best explanation for the CMBR.

Robert Woodrow Wilson

 Robert Woodrow Wilson is an American astronomer, 1978 Nobel laureate in physics, who with Arno Allan Penzias in 1964 discovered the cosmic microwave background radiation (CMB). The award purse was also shared with a 3rd scientist, Pyotr Leonidovich Kapitsa for unrelated work.

Born	10 January 1936 (age 74), Houston, Texas, USA
Nationality	United States
Fields	Physics
Alma mater	Rice University
Known for	Cosmic microwave background radiation
Notable awards	Nobel Prize in Physics (1978)

COSMIC MICROWAVE BACKGROUND RADIATION (1978)

The cosmic microwave background is isotropic to roughly one part in 100,000: the root mean square variations are only 18 µK. *The Far-Infrared Absolute Spectrophotometer* (*FIRAS*) instrument on the NASA Cosmic Background Explorer (COBE) satellite has carefully measured the spectrum of the cosmic microwave background. The FIRAS project members compared the CMB with an internal reference black body and the spectra agreed to within the experimental error. They concluded that any deviations from the black body form that might still remain undetected in the CMB spectrum over the wavelength range from 0.5 to 5 mm must have a weighted rms value of at most 50 parts per million (0.005%) of the CMB peak brightness. This made the CMB spectrum the most precisely measured black body spectrum in nature.

The cosmic microwave background is perhaps the main prediction of the Big Bang model. In addition, Inflationary

Cosmology predicts that after about 10^{-37} seconds the nascent universe underwent exponential growth that smoothed out nearly all in homogeneities. This was followed by symmetry breaking; a type of phase transition that set the fundamental forces and elementary particles in their present form. After 10^{-6} seconds, the early universe was made up of a hot plasma of photons, electrons, and baryons. The photons were constantly interacting with the plasma through Thomson scattering. As the universe expanded, adiabatic cooling caused the plasma to cool until it became favorable for electrons to combine with protons and form hydrogen atoms. This recombination event happened at around 3000 K or when the universe was approximately 379,000 years old. At this point, the photons scattered off the now electrically neutral atoms and began to travel freely through space, resulting in the decoupling of matter and radiation.

The colour temperature of the photons has continued to diminish ever since; now down to 2.725 K, their temperature will continue to drop as the universe expands. According to the Big Bang model, the radiation from the sky we measure today comes from a spherical surface called *the surface of last scattering*. This represents the collection of spots in space at which the decoupling event is believed to have occurred, less than 400,000 years after the Big Bang, and at a point in time such that the photons from that distance have just reached observers. The estimated age of the Universe is 13.75 billion years. However, because the Universe has continued expanding since that time, the comoving distance from the Earth to the edge of the observable universe is now at least 46.5 billion light years.

The Big Bang theory suggests that the cosmic microwave background fills all of observable space, and that most of the radiation energy in the universe is in the cosmic microwave background, which makes up a fraction of roughly 6×10^{-5} of the total density of the universe.

Sheldon Lee Glashow

 Sheldon Lee Glashow is a Nobel Prize winning American physicist. He is the Metcalf Professor of Mathematics and Physics at Boston University. Sheldon Lee Glashow was born in New York City to Jewish immigrants from Russia. He attended the Bronx High School of Science in New York City.

In 1961, Glashow extended electroweak unification models due to Schwinger by including a short range neutral current, the Z0. The resulting symmetry structure that Glashow proposed, SU(2) × U(1), forms the basis of the accepted theory of the electroweak interactions. For this discovery, Glashow along with Steven Weinberg and Abdus Salam, was awarded the 1979 Nobel Prize in Physics.

Born	5 December 1932 (age 77), New York City, New York, USA
Nationality	United States
Fields	Physics
Institutions	Boston University, Harvard University
Alma mater	Cornell University, Harvard University
Known for	Electroweak theory, Criticism of superstring theory
Notable awards	Nobel Prize in Physics (1979)

ELECTROWEAK INTERACTION (1979)

In particle physics, the electroweak interaction is the unified description of two of the four fundamental interactions of nature: electromagnetism and the weak interaction. Although these two forces appear very different at everyday low energies, the theory models them as two different aspects of the same force. Above the unification energy, on the order of 100 Gev, they would merge into a single electroweak force. Thus if the universe is hot enough then the electromagnetic force and weak force will merge into a combined electroweak force.

For contributions to the unification of the weak and electromagnetic interaction between elementary particles, Abdus Salam, Sheldon Glashow and Steven Weinberg were awarded the Nobel Prize in Physics in 1979. The existence of the electroweak interactions was experimentally established in two stages: the first being the discovery of neutral currents in neutrino scattering by the Gargamelle collaboration in 1973, and the second in 1983 by the UA1 and the UA2 collaborations that involved the discovery of the W and Z gauge bosons in proton-antiproton collisions at the converted Super Proton Synchrotron.

Superstring theory is an attempt to explain all of the particles and fundamental forces of nature in one theory by modelling them as vibrations of tiny supersymmetric strings. Superstring theory is a shorthand for supersymmetric string theory because unlike bosonic string theory, it is the version of string theory that incorporates fermions and supersymmetry. Here, the word super has the meaning of super-symmetry.

The deepest problem in theoretical physics is harmonizing the theory of general relativity, which describes gravitation and applies to large-scale structures, with quantum mechanics, which describes the other three fundamental forces acting on the atomic scale.

The development of a quantum field theory of a force invariably results in infinite probabilities. Physicists have developed mathematical techniques to eliminate these infinities which work for three of the four fundamental forces–electromagnetic, strong nuclear and weak nuclear forces–but not for gravity. The development of a quantum theory of gravity must therefore come about by different means than those used for the other forces.

The basic idea is that the fundamental constituents of reality are strings of the Planck length (about 10^{-33} cm) which vibrate at resonant frequencies. Every string in theory has a unique resonance, or harmonic. Different harmonics determine different fundamental forces. The tension in a string is on the order of the Planck force (10^{44} newtons).

Mohammad Abdus Salam

 Mohammad Abdus Salam, was a Pakistani theoretical physicist, astrophysicist and Nobel laureate in Physics for his work in Electro-Weak Theory. Salam, Sheldon Glashow and Steven Weinberg shared the prize for this discovery.

Salam holds the distinction of being the first Pakistani and the first Muslim Nobel Laureate to receive the prize in the Sciences. Even today, Salam was considered one of the most influential scientists and physicists in his country.

Born	29 January 1926, Jhang, Punjab, British Raj
Died	21 November 1996 (age 70), Oxford, England, United Kingdom
Citizenship	Pakistani
Nationality	Pakistani
Fields	Theoretical Physics
Institutions	Pakistan Atomic Energy Commission (PAEC), Space and Upper Atmosphere Research Commission (SUPARCO), Punjab University, Imperial College, London, Government College, University of Cambridge, International Centre for Theoretical Physics (ICTP), COMSATSTWAS, Edward Bouchet Abdus Salam Institute.
Alma mater	University of the Punjab, Government College, St John's College, Cambridge
Doctoral advisor	Nicholas Kemmer Paul Matthews
Doctoral students	Michael Duff, Walter Gilbert, John Moffat, Yuval Ne'eman, John Polkinghorne, Raziuddin Siddiqui, Riazuddin, Fayyazuddin, Masud Ahmad, Ghulam Murtaza, Faheem Hussain
Known for	Electroweak theory, Pati-Salam model, Pakistan's atomic program Pakistan's space program
Notable awards	Nobel Prize in Physics (1979), Smith's Prize, Adams Prize, Nishan-e-Imtiaz (1979), Sitara-e-Pakistan (1959)

ELECTROWEAK INTERACTION FORMULATION (1979)

Mathematically, the unification is accomplished under an $SU(2) \times U(1)$ gauge group. The corresponding gauge bosons are the photon of electromagnetism and the W and Z bosons of the weak force. In the Standard Model, the weak gauge bosons get their mass from the spontaneous symmetry breaking of the electroweak symmetry from $SU(2) \times U(1)_Y$ to $U(1)_{em}$, caused by the Higgs mechanism The subscripts are used to indicate that these are different copies of $U(1)$; the generator of $U(1)_{em}$ is given by $Q = Y/2 + I_3$, where Y is the generator of $U(1)_Y$ (called the weak hypercharge), and I_3 is one of the $SU(2)$ generators (a component of weak isospin). The distinction between electromagnetism and the weak force arises because there is a (nontrivial) linear combination of Y and I_3 that vanishes for the Higgs boson (it is an eigenstate of both Y and I_3, so the coefficients may be taken as $-I_3$ and Y): $U(1)_{em}$ is defined to be the group generated by this linear combination, and is unbroken because it does not interact with the Higgs.

PATI-SALAM MODEL

In physics, the Pati-Salam model is a Grand Unification Theory (GUT) was proposed in 1974 by nobel laureate Abdus Salam and Jogesh Pati. The unification is based on there being four quark colour charges, dubbed red, green, blue and violet (or lilac), instead of the conventional three, with the new violet quark being identified with the leptons. The model also has Left-right symmetry and predicts the existence of a high energy right handed weak interaction with heavy W' and Z' bosons . Originally the fourth colour was labeled lilac to alliterate with lepton Pati-Salam is a mainstream theory and a viable alternative to the Georgi-Glashow SU(5) unification. It can be embedded within an SO(10) unification model (as can SU(5)).

Steven Weinberg

Steven Weinberg is an American physicist and Nobel laureate in Physics for his contributions with Abdus Salam and Sheldon Glashow to the unification of the weak force and electromagnetic interaction between elementary particles.

It is of special importance that in 1979 he pioneered the modern view on the renormalization aspect of quantum field theory that considers all quantum field theories as effective field theories and changed completely the viewpoint of previous work that a sensible quantum field theory must be renormalizable. This approach allowed the development of effective theory of quantum gravity, low energy QCD, heavy quark effective field theory and other developments, and it is a topic of considerable interest in current research.

Born	3 May 1933 (age 77), New York City, New York, USA
Residence	United States
Nationality	United States
Fields	Physics
Institutions	MIT, Harvard University, University of Texas at Austin
Alma mater	Cornell University, Princeton University
Doctoral advisor	Sam Treiman
Doctoral students	Orlando Alvarez, Claude Bernard, Lay Nam Chang, Bob Holdom, Ubirajara van Kolck, Rafael Lopez-Mobilia, John Preskill, Fernando Quevedo, Mark G. Raizen, Scott Willenbrock
Known for	Electromagnetism and Weak Force unification, Weinberg-Witten theorem
Notable awards	Nobel Prize in Physics (1979)

WEAK INTERACTION (1979)

The weak interaction is one of the four fundamental interactions of nature, along with strong interaction, electromagnetic force, and gravitation. In the Standard Model of particle physics, it is due to the exchange of the heavy W and Z bosons. Its most familiar effect is beta decay and the associated radioactivity. The word *weak* derives from the fact that the typical field strength is 10^{-11} times the strength of the electromagnetic force and some 10^{-13} times that of the strong force, when forces are compared between particles interacting in more than one way. The weak interaction affects all left handed leptons and quarks. Other than gravity, it is the only force affecting neutrinos. The weak interaction is unique in a number of respects:

1.   It is the only interaction capable of changing flavour.
2.   It is the only interaction which violates parity symmetry P (because it almost exclusively acts on left-handed particles). It is also the only one which violates CP (CP Symmetry).
3.   It is mediated by *massive* gauge bosons. This unusual feature is explained in the Standard Model by the Higgs mechanism.

Due to the large mass of the weak interaction's carrier particles (about 90 Gev/c^2), their mean life is about 3×10^{-25} seconds.

Since the weak interaction is both very weak and very short range, its most noticeable effect is due to its other unique feature: flavor changing. Consider a neutron (quark content: *udd*, or one up quark and two down quarks). Although the neutron is heavier than its sister nucleon, the proton (quark content *uud*), it cannot decay into a proton without changing the flavor of one of its down quarks. Neither the strong interaction nor electromagnetism allow flavor changing, so this must proceed by weak decay. In this process, a down quark in the neutron changes into an up quark by emitting a W^- boson, which then breaks up into a high-energy electron and an electron antineutrino. Since high-energy electrons are beta radiation, this is called a beta decay.

James Watson Cronin

 James Watson Cronin an American nuclear physicist. Cronin was born in Chicago, Illinois and attended Southern Methodist University in Dallas, Texas. Cronin and co-researcher Val Logsdon Fitch were awarded the 1980 Nobel Prize in Physics for a 1964 experiment that proved that certain subatomic reactions do not adhere to fundamental symmetry principles.

Specifically, they proved, by examining the decay of kaons, that a reaction run in reverse does not merely retrace the path of the original reaction, which showed that the interactions of subatomic particles are not indifferent to time. Thus the phenomenon of CP violation was discovered.

Born	29 September 1931 (age 78) Chicago, Illinois, USA
Nationality	United States
Fields	Physics
Institutions	University of Chicago
Alma mater	Southern Methodist University, University of Chicago
Known for	Nuclear physics
Notable awards	Nobel Prize in Physics (1980), John Price Wetherill Medal

CP VIOLATION (1980)

CP is the product of two symmetries: C for charge conjugation, which transforms a particle into its antiparticle, and P for parity, which creates the mirror image of a physical system. The strong interaction and electromagnetic interaction seem to be invariant under the combined CP transformation operation, but this symmetry is slightly violated during certain types of weak decay. Historically, CP symmetry was proposed to restore order after the discovery of parity violation in the 1950s.

The idea behind parity symmetry is that the equations of particle physics are invariant under mirror inversion. This leads to the prediction that the mirror image of a reaction (such as a chemical reaction or radioactive decay) occurs at the same rate as the original reaction. Parity symmetry appears to be valid for all reactions involving electromagnetism and strong interactions. Until 1956, parity conservation was believed to be one of the fundamental geometric conservation laws (along with conservation of energy and conservation of momentum). However, in 1956 a careful critical review of the existing experimental data by theoretical physicists Tsung-Dao Lee and Chen Ning Yang revealed that while parity conservation had been verified in decays by the strong or electromagnetic interactions, it was untested in the weak interaction. They proposed several possible direct experimental tests. The first test based on beta decay of Cobalt-60 nuclei was carried out in 1956 by a group led by Chien-Shiung Wu, and demonstrated conclusively that weak interactions violate the P symmetry or, as the analogy goes, some reactions did not occur as often as their mirror image.

Overall, the symmetry of a quantum mechanical system can be restored if another symmetry S can be found such that the combined symmetry PS remains unbroken. This rather subtle point about the structure of Hilbert space was realized shortly after the discovery of P violation, and it was proposed that charge conjugation was the desired symmetry to restore order.

Simply speaking, charge conjugation is a simple symmetry between particles and antiparticles, and so CP symmetry was proposed in 1957 by Lev Landau as the true symmetry between matter and antimatter. In other words a process in which all particles are exchanged with their antiparticles was assumed to be equivalent to the mirror image of the original process.

Val Logsdon Fitch

Val Logsdon Fitch is an American nuclear physicist. A native of Merriman, Nebraska, he graduated from Gordon High School and attended Chadron State College for three years before being drafted into the U.S. army in 1943. In World War II, he worked on the Manhattan Project in Los Alamos.

He is a member of the faculty at Princeton University. Fitch and co-researcher James Watson Cronin were awarded the 1980 Nobel Prize in Physics for a 1964 experiment using the Alternating Gradient Synchrotron at Brookhaven National Laboratory that proved that certain subatomic reactions do not adhere to fundamental symmetry principles. Specifically, they proved, by examining the decay of K-mesons, that a reaction run in reverse does not merely retrace the path of the original reaction, which showed that the reactions of subatomic particles are not indifferent to time. Thus the phenomenon of CP violation was discovered.

Born	10 March 1923 (age 87), Merriman, Nebraska
Fields	Particle physics
Institutions	Princeton
Alma mater	Columbia, McGill University
Known for	Discovery of CP-violation
Notable awards	John Price Wetherill Medal (1976), Nobel Prize in Physics (1980)

INDIRECT CP VIOLATION (1980)

In 1964, James Cronin, Val Fitch with coworkers provided clear evidence that CP symmetry could be broken, too, winning them the 1980 Nobel Prize. This discovery showed that weak interactions violate not only the charge-conjugation symmetry C between particles and antiparticles and the P or parity, but also their combination. The discovery shocked particle physics and opened the door to questions still at the core of particle physics and of cosmology today. The lack of an exact CP symmetry, but also the fact that it is so nearly a symmetry, created a great puzzle.

Only a weaker version of the symmetry could be preserved by physical phenomena, which was CPT symmetry. Besides C and P, there is a third operation, time reversal (T), which corresponds to reversal of motion. Invariance under time reversal implies that whenever a motion is allowed by the laws of physics, the reversed motion is also an allowed one. The combination of CPT is thought to constitute an exact symmetry of all types of fundamental interactions. Because of the CPT symmetry, a violation of the CP symmetry is equivalent to a violation of the T symmetry. CP violation implied nonconservation of T, provided that the long-held CPT theorem was valid. In this theorem, regarded as one of the basic principles of quantum field theory, charge conjugation, parity, and time reversal are applied together.

Direct CP violation

The kind of CP violation discovered in 1964 was linked to the fact that neutral kaons can transform into their antiparticles and vice versa, but such transformation does not occur with exactly the same probability in both directions; this is called *indirect* CP violation. Despite many searches, no other manifestation of CP violation was discovered until the 1990s, when the NA31 experiment at CERN suggested evidence for CP violation in the decay process of the very same neutral kaons. The observation was somewhat controversial, and final proof for it came in 1999 from the KTeV experiment at Fermi lab and the NA48 experiment at CERN.

In 2001, a new generation of experiments, including the BaBar Experiment at the Stanford Linear Accelerator Center (SLAC) and the Belle Experiment at the High Energy Accelerator Research Organization (KEK) in Japan, observed direct CP violation in a different sector of particle physics, namely in decays of the B mesons. By now a large number of CP violation processes in B meson decays have been discovered. Before these B factory experiments, it was a logical possibility that all CP violation was confined to kaon physics. However, this raised the question of why it's *not* extended to the strong force, and furthermore, why this is not predicted in the unextended Standard Model, despite the model being undeniably accurate with normal phenomena.

Nicolaas Bloembergen

Nicolaas Bloembergen is a Dutch American physicist and Nobel laureate. He received his PhD degree from University of Leiden in 1948; while pursuing his PhD at Harvard, Bloembergen also worked part-time as a graduate research assistant for Edward Mills Purcell at the MIT Radiation Laboratory.

He became a professor at Harvard University in Applied Physics. He was awarded the Lorentz Medal in 1978. Bloembergen shared the 1981 Nobel Prize in Physics with Arthur Schawlow and Kai Siegbahn for their work in laser spectroscopy. Bloembergen and Schawlow investigated properties of matter undetectable without lasers. He had earlier modified the maser of Charles Townes.

Born	11 March 1920 (age 90), Dordrecht, Netherlands
Residence	United States
Citizenship	Netherlands, United States
Fields	Applied physics
Institutions	University of Arizona
Alma mater	Leiden, University of Utrecht
Doctoral advisor	Edward Purcell
Known for	Laser spectroscopy
Notable awards	Nobel Prize in Physics (1981), Lorentz Medal (1978), IEEE Medal of Honor

SPECTROSCOPY (1981)

Spectroscopy was originally the study of the interaction between radiation and matter as a function of wavelength (λ). In fact, historically, spectroscopy referred to the use of visible light dispersed according to its wavelength, e.g. by a prism. Later the concept was expanded greatly to comprise any measurement of a quantity as a function of either wavelength or frequency. Thus, it also can refer to a response to an alternating field or varying frequency (ν). A further extension of the scope of the definition added energy (E) as a variable, once the very close relationship $E = h\nu$ for photons was realized (h is the Planck constant). A plot of the response as a function of wavelength or more common frequency referred to as a spectrum.

Spectrometry is the spectroscopic technique used to assess the concentration or amount of a given chemical (atomic, molecular, or ionic) species. In this case, the instrument that performs such measurements is a spectrometer, spectrophotometer, or spectrograph.

Spectroscopy/spectrometry is often used in physical and analytical chemistry for the identification of substances through the spectrum emitted from or absorbed by them.

Spectroscopy/spectrometry is also heavily used in astronomy and remote sensing. Most large telescopes have spectrometers, which are used either to measure the chemical composition and physical properties of astronomical objects or to measure their velocities from the Doppler shift of their spectral lines.

Absorption spectroscopy usually implies having a tunable frequency source and producing a plot of absorption as a function of frequency. This was not feasible with lasers until the advent of the dye lasers which can be tuned over a nearly continuous range of frequencies.

Laser spectroscopy has led to advances in the precision with which spectral line frequencies can be measured, and this has fundamental significance for our understanding of basic atomic processes. This precision has been obtained by passing two laser beams through the absorption sample in opposite directions, selectively triggering absorption only in those atoms that have a zero velocity component in the direction of the beams. This effectively eliminates the Doppler broading of spectral lines from the distribution of atomic velocities present in the sample.

Arthur Leonard Schawlow

Arthur Leonard Schawlow was an American physicist. He is best remembered for his work on lasers, for which he was awarded a 1981 Nobel Prize.

Although his research focused on optics, in particular, lasers and their use in spectroscopy, he also pursued investigations in the areas of superconductivity and nuclear resonance. Schawlow shared the 1981 Nobel Prize in Physics with Nicolaas Bloembergen and Kai Siegbahn for their contributions to the development of laser spectroscopy.

Born	5 May 1921, Mount Vernon, New York
Died	28 April 1999 (age 77), Palo Alto, California
Nationality	United States
Fields	Physics
Institutions	Bell Labs, Columbia University, Stanford University
Alma mater	University of Toronto
Doctoral advisor	Malcolm Crawford
Known for	Laser spectroscopy
Notable awards	Nobel Prize in Physics (1981)

LASER (1981)

Light Amplification by Stimulated Emission of Radiation (LASER or laser) is a mechanism for emitting electromagnetic radiation, typically light or visible light, via the process of stimulated emission. The emitted laser light is (usually) a spatially coherent, narrow low-divergence beam, that can be manipulated with lenses. In laser technology, coherent light denotes a light source that produces (emits) light of in-step waves of identical frequency, phase, and polarization. The laser's beam of coherent light differentiates it from light sources that emit incoherent light beams, of random phase varying with time and position. Laser light is generally a narrow-wavelength electromagnetic spectrum monochromatic light; yet, there are lasers that emit a broad spectrum of light, or emit different wavelengths of light simultaneously.

The word laser originally was the upper-case LASER, the acronym from Light Amplification by Stimulated Emission of Radiation, wherein light broadly denotes electromagnetic radiation of any frequency, not only the visible spectrum; hence infrared laser, ultraviolet laser, X-ray laser, et cetera. Because the microwave predecessor of the laser, the maser, was developed first, devices that emit microwave and radio frequencies are denoted "masers". In the early technical literature, especially in that of the Bell Telephone Laboratories researchers, the laser was also called optical maser, a currently uncommon term. Moreover, since 1998, Bell Laboratories adopted the laser usage. Linguistically, the back-formation verb to lase means to produce laser light and to apply laser light to. The word laser sometimes is used in an extended sense to describe a non-laser-light technology, e.g. a coherent-state atom source is an atom laser.

The emitted laser light is notable for its high degree of spatial and temporal coherence, unattainable using other technologies.

Spatial coherence, typically is expressed through the output being a narrow beam which is diffraction-limited, often a so-called pencil beam. Laser beams can be focused to very tiny spots, achieving a very high irradiance. Or they can be launched into a beam of very low divergence in order to concentrate their power at a large distance.

Temporal coherence implies a polarized wave at a single frequency whose phase is correlated over a relatively large distance along the beam. A beam produced by a thermal or other incoherent light source has an instantaneous amplitude and phase which vary randomly with respect to time and position, and thus a very short coherence length.

Kai Manne Börje Siegbahn

 Kai Manne Börje Siegbahn was a Swedish physicist. He was born in Lund, Sweden, and his father Manne Siegbahn also won the Nobel Prize in Physics, in 1924. Siegbahn earned his doctorate at the University of Stockholm in 1944.

He was professor at the Royal Institute of Technology 1951–1954, and then professor of experimental physics at Uppsala University 1954–1984, which was the same chair his father had held. He shared the 1981 Nobel Prize in Physics with Nicolaas Bloembergen and Arthur Schawlow for their work in spectroscopy. Siegbahn obtained the Nobel Prize for developing the method of Electron Spectroscopy for Chemical Analysis (ESCA), now usually described as X-ray photoelectron spectroscopy (XPS).

Born	20 April 1918, Lund, Sweden
Died	20 July 2007 (age 89), Angelholm, Sweden
Nationality	Sweden
Fields	Physics
Institutions	University of Stockholm, University of Uppsala
Alma mater	University of Stockholm
Known for	high-resolution electron spectroscopy
Notable awards	Nobel Prize in Physics (1981)

ELECTRON SPECTROSCOPY (1981)

Electron spectroscopy is an analytical technique to study the electronic structure and its dynamics in atoms and molecules. In general an excitation source such as X-rays, electrons or synchrotron radiation will eject an electron from an inner-shell orbital of an atom. Detecting photoelectrons that are ejected by X-rays is called X-ray photoelectron spectroscopy (XPS) or electron spectroscopy for chemical analysis (ESCA). Detecting electrons that are ejected from higher orbital's to conserve energy during electron transitions is called Auger Electron Spectroscopy (AES).

Experimental applications include high-resolution measurements on the intensity and angular distributions of emitted electrons as well as on the total and partial ion yields. Ejected electrons can escape only from a depth of approximately 3 nanometers or less, making electron spectroscopy most useful to study surfaces of solid materials. Depth profiling is accomplished by combining an electron spectroscopy with a sputtering source that removes surface layers.

Synchrotron radiation research work has been carried out at the MAX Laboratory in Lund, Sweden, Elettra Storage Ring in Trieste, Italy, and at ALS in Berkeley, CA.

X-ray photoelectron spectroscopy (XPS) is a quantitative spectroscopic technique that measures the elemental composition, empirical formula, chemical state and electronic state of the elements that exist within a material. XPS spectra are obtained by irradiating a material with a beam of X-rays while simultaneously measuring the kinetic energy and number of electrons that escape from the top 1 to 10 nm of the material being analyzed. XPS requires ultra high vacuum (UHV) conditions.

XPS is a surface chemical analysis technique that can be used to analyze the surface chemistry of a material in it as received state, or after some treatment, for example: fracturing, cutting or scraping in air or UHV to expose the bulk chemistry, ion beam etching to clean off some of the surface contamination, exposure to heat to study the changes due to heating, exposure to reactive gases or solutions, exposure to ion beam implant, exposure to ultraviolet light.

Kenneth Geddes Wilson

Kenneth Geddes Wilson is an American theoretical physicist. He joined Cornell University in 1963 in the Department of Physics as a junior faculty member, becoming a full professor in 1970. In 1974, he became the James A. Weeks Professor of Physics at Cornell.

He was a co-winner of the Wolf Prize in physics in 1980, together with Michael E. Fisher and Leo Kadanoff. He was awarded the 1982 Nobel Prize in Physics for his seminal approach, combining quantum field theory and the statistical theory of critical phenomena of second-order phase transitions, i.e. for his constructive theory of the renormalization group. In this theory he gave not only important, and even numerical, insights to the field of critical statics and dynamics in statistical physics, but indirectly also basic answers to the question: What is quantum field theory? and What does renormalization mean? He also gave a constructive answer to another important renormalization problem from solid-state physics, the Kondo effect.

Born	8 June 1936 (age 73), Massachusetts, USA
Nationality	United States
Fields	Theoretical physics
Institutions	Cornell University
Alma mater	Caltech
Doctoral advisor	Murray Gell-Mann
Doctoral students	Roman Jackiw, Steve Shenker, Michael Peskin
Known for	Phase transitions, Wilson loops
Notable awards	Wolf Prize in physics (1980), Nobel Prize in Physics (1982)

PHASE TRANSITION (1982)

A phase transition is the transformation of a thermodynamic system from one phase or state of matter to another.

A phase of a thermodynamic system and the states of matter have essentially uniform physical properties. During a phase transition of a given medium certain properties of the medium change, often discontinuously, as a result of some external condition, such as temperature, pressure, and others. For example, a liquid may become gas upon heating to the boiling point, resulting in an abrupt change in volume. The measurement of the external conditions at which the transformation occurs, is termed the *phase transition point*.

Phase transitions are common occurrences observed in nature and many engineering techniques exploit certain types of phase transition.

The term is most commonly used to describe transitions between solid, liquid and gaseous states of matter, in rare cases including plasma.

Wilson loop

In gauge theory, a Wilson loop (named after Kenneth G. Wilson) is a gauge-invariant observable obtained from the holonomy of the gauge connection around a given loop. In the classical theory, the collection of all Wilson loops contains sufficient information to reconstruct the gauge connection, up to gauge transformation.

In quantum field theory, the definition of Wilson loop observables as *bona fide* operators on Fock space (actually, Haag's theorem states that Fock space does not exist for interacting QFTs) is a mathematically delicate problem and requires regularization, usually by equipping each loop with a *framing*. The action of Wilson loop operators has the interpretation of creating an elementary excitation of the quantum field which is localized on the loop. In this way, Faraday's flux tubes become elementary excitations of the quantum electromagnetic field.

Subrahmanyan Chandrasekhar

 Subrahmanyan Chandrasekhar, was an Indian-American astrophysicist. He was a Nobel laureate in physics along with William Alfred Fowler for their work in the theoretical structure and evolution of stars.

He was the nephew of Indian Nobel Laureate Sir C. V. Raman. Chandrasekhar served on the University of Chicago faculty from 1937 until his death in 1995 at the age of 84. He became a naturalized citizen of the United States in 1953.

Born	19 October 1910, Lahore, Punjab, British India
Died	21 August 1995 (age 84), Chicago, Illinois, United States
Nationality	British India (1910–1947), India (1947–1953), United States (1953–1995)
Fields	Astrophysics
Institutions	University of Chicago, University of Cambridge
Alma mater	Trinity College, Cambridge, Presidency College, Madras
Doctoral advisor	R.H. Fowler
Doctoral students	Donald Edward Osterbrock, Roland Winston
Known for	Chandrasekhar limit
Notable awards	Nobel Prize in Physics (1983), Copley Medal (1984), National Medal of Science (1966), Padma Vibhushan (1968)

CHANDRASEKHAR LIMIT (1983)

The Chandrasekhar limit, limits the mass of bodies made from electron-degenerate matter, a dense form of matter which consists of nuclei immersed in a gas of electrons. The limit is the maximum nonrotating mass which can be supported against gravitational collapse by electron degeneracy pressure.

It is named after the Indian astrophysicist Subrahmanyan Chandrasekhar, and is commonly given as being about 1.4 solar masses. As white dwarfs are composed of electron-degenerate matter, no nonrotating white dwarf can be heavier than the Chandrasekhar limit. The Chandrasekhar limit is analogous to the Tolman-Oppenheimer-Volkoff limit for Neutron Stars.

Stars produce energy through nuclear fusion, producing heavier elements from lighter ones. The heat generated from these reactions prevents gravitational collapse of the star. Over time, the star builds up a central core which consists of elements which the temperature at the center of the star is not sufficient to fuse. For main-sequence stars with a mass below approximately 8 solar masses, the mass of this core will remain below the Chandrasekhar limit, and they will eventually lose mass (as planetary nebulae) until only the core, which becomes a white dwarf, remains. Stars with higher mass will develop a degenerate core whose mass will grow until it exceeds the limit. At this point the star will explode in a core-collapse supernova, leaving behind either a neutron star or a black hole.

Computed values for the limit will vary depending on the approximations used, the nuclear composition of the mass, and the temperature gives a value of

$$M_{limit} = \frac{w_3^0 \sqrt{3\pi}}{2} \left(\frac{\hbar c}{G} \right)^{3/2} \frac{1}{\left(\mu_e m_H \right)^2}$$

here, μ_e is the average molecular weight per electron, m_H is the mass of the hydrogen atom, and $w_3^0 \approx 2.018236$ is a constant connected with the solution to the Lane-Emden equation. Numerically, this value is approximately $(2/\mu_e)^2 \times 2.85 \times 10^{30}$ kg, or 1.43 $(2/\mu_e)^2$ $M_\odot$, where $M_\odot = 1.989 \times 10^{30}$ kg is the standard solar mass. As $\sqrt{\hbar c/G}$ is the Planck mass, $M_{Pl} \approx 2.176 \times 10^{-8}$ kg, the limit is of the order of $\dfrac{M_{Pl}^3}{m_H^2}$.

William Alfred Willie Fowler

 William Alfred Willie Fowler was an American astrophysicist. His seminal paper *Synthesis of the Elements in Stars* (*Reviews of Modern Physics*, vol. 29, Issue 4, pp. 547–650), co-authored with E. Margaret Burbidge, Geoffrey Burbidge, and Fred Hoyle, was published in 1957.

The paper explained how the abundances of essentially all but the lightest chemical elements could be explained by the process of nucleosynthesis in stars. It is widely known as B^2FH. Fowler won the Henry Norris Russell Lectureship of the American Astronomical Society in 1963, the Vetlesen Prize in 1973, the Eddington Medal in 1978, the Bruce Medal of the Astronomical Society of the Pacific in 1979, and the Nobel Prize for Physics in 1983 for his theoretical and experimental studies of the nuclear reactions of importance in the formation of the chemical elements in the universe (shared with Subrahmanyan Chandrasekhar).

Born	9 August 1911, Pittsburgh, Pennsylvania
Died	14 March 1995 (age 83), Pasadena, California
Doctoral advisor	Charles Christian Lauritsen
Doctoral students	George Fuller, Donald Clayton
Notable awards	Nobel Prize in Physics (1983)

NUCLEAR REACTION (1983)

In nuclear physics and nuclear chemistry, a nuclear reaction is the process in which two nuclei or nuclear particles collide to produce products different from the initial particles. In principle a reaction can involve more than three particles colliding, but because the probability of three or more nuclei to meet at the same time at the same place is much less than for two nuclei, such an event is exceptionally rare. While the transformation is spontaneous in the case of radioactive decay, it is initiated by a particle in the case of a nuclear reaction. If the particles collide and separate without changing, the process is called an elastic collision rather than a reaction.

$^{6}_{3}Li$ and deuterium $^{2}_{1}H$ react to form the highly excited intermediate nucleus 84Be which then decays immediately into two alpha particles. Protons are symbolically represented by red spheres, and neutrons by blue spheres.

$$^{6}_{3}Li + ^{2}_{1}H \rightarrow ^{4}_{2}He + ?$$

To balance the equation above, the second nucleus to the right must have atomic number 2 and mass number 4; it is therefore also helium-4. The complete equation therefore reads:

$$^{6}_{3}Li + ^{2}_{1}H \rightarrow ^{4}_{3}He + ^{4}_{2}He$$

or more simply:

$$^{6}_{3}Li + ^{2}_{1}H \rightarrow ^{4}_{2}He$$

Natural nuclear reactions occur in the interaction between cosmic rays and matter. On Earth, nuclear reactions initiated by radioactive decay produce nucleogenic isotopes rather than the radiogenic daughter isotopes of simple decay. Unlike radioactive decay, nuclear reactions can be employed artificially to obtain nuclear energy, at an adjustable rate, on demand. Perhaps the most notable nuclear reactions are the nuclear chain reaction that produces nuclear fission, and the nuclear fusion reactions that power the energy production of the Sun and stars. Both of these types of reactions are employed in nuclear weapons.

Instead of using the full equations as shown in the previous section, in many situations a compact notation is used to describe nuclear reactions. This is A (b, c) D, which is equivalent to A + b gives c + D. Common light particles are often abbreviated in this shorthand, typically p for proton, n for neutron, d for deuteron, α representing an alpha particle or helium-4, β for beta particle or electron, γ for gamma photon, etc. The reaction above would be written as Li-6 (d, α) α.

Carlo Rubbia

Carlo Rubbia is an Italian particle physicist and inventor who won the Nobel Prize in Physics in 1984 for the work leading to the discovery of the W and Z particles at CERN.

To achieve energies high enough to create these particles, Rubbia, together with David Cline and Peter McIntyre, proposed a radically new particle accelerator design. They proposed to use a beam of protons and a beam of antiprotons, their antimatter twins, counter rotating in the vacuum pipe of the accelerator and colliding head-on. As a result, scientists had to develop a number of techniques for creating and handling intense beams of antiprotons. Many of the new experimental methods were developed by Simon van der Meer, who shared the 1984 Nobel Prize for Physics with Rubbia.

Born	31 March 1934 (age 76), Gorizia, Friuli-Venezia Giulia, Italy
Nationality	Italian
Fields	Physics
Known for	Discovery of W and Z bosons
Notable awards	Nobel Prize in Physics (1984)

W AND Z BOSONS (1984)

The W and Z^0 bosons are the elementary particles that mediate the weak force. Their discovery was a major success for what is now called the Standard Model of particle physics.

The W particle is named after the weak nuclear force. The physicist Steven Weinberg named the additional particle the Z particle, giving no explanation. It has been speculated that the Z particle was semi-humorously given its name because it was said to be the last particle to need to be discovered. Another explanation for the name is that the Z particle derives its name from it having zero electric charge (i.e. being an uncharged particle).

There are two types of W bosons; the W^+ with an electric charge of +1 e and its antiparticle, the W^- with an electric charge of −1 e. The Z boson (or Z particle) is electrically neutral, and it is its own antiparticle. All three of these particles are very short-lived with a half-life of about 3×10^{-25} s. In practice, they can be considered to be virtual particles.

These bosons are among the heavyweights of the elementary particles. With masses of 80.4 Gev/c^2 and 91.2 Gev/c^2, respectively, the W and Z particles are almost 100 times as massive as the proton—heavier than entire atoms of iron. The masses of these bosons are significant because they act as the force carriers of a quite short-range fundamental force: their high masses thus limit the range of the weak nuclear force. By way of contrast, the electromagnetic force, has an infinite range because its force carrier, the photon, has zero rest mass. The significant mass of the pion limits the range of the strong nuclear force.

All three types of these W and Z particles have a particle spin values of plus or minus one. The emission of a W^+ or W^- boson either raises or lowers the electric charge of the emitting particle by one unit, and also alters the spin by one unit. At the same time, the admission of absorption of a W boson can change the type of the particle - for example changing a strange quark into an up quark. The neutral Z boson obviously cannot change the electric charge of any particle, nor can it change any other of the so-called charges (such as strangeness, baryon number, charm, etc.). The emission or absorption of a Z particle can only change the spin, momentum, and energy of the other particle.

Simon van der Meer

 Simon van der Meer is a Dutch accelerator physicist who won the Nobel Prize in Physics in 1984 for his contributions to the project which led to the discovery of the W and Z particles at CERN.

Born	24 November 1925 (age 84), The Hague, The Netherlands
Nationality	Dutch
Fields	Physics
Known for	Stochastic cooling
Notable awards	Nobel Prize in Physics (1984)

Simon van der Meer was born and grow up in The Hague, finishing his secondary education during the German occupation of the Netherlands. He studied Technical Physics at the Delft University of Technology, and received an engineer's degree in 1952. After worked for Philips for a few years, in 1956 he joined CERN, where he stayed until his retirement in 1990.

Van der Meer invented the technique of stochastic cooling of particle beams. This technique was used to accumulate intense beams of protons and antiprotons in the Super Proton Synchrotron at CERN, which allowed the UA1 experiment, led by Carlo Rubbia, to produce W and Z bosons through 500 Gev proton-antiproton collisions in early 1983. The W and Z bosons had been theoretically predicted some years earlier, and their experimental discovery was considered a significant success for CERN. Van der Meer and Rubbia shared the 1984 Nobel Prize for their decisive contributions to the project.

Van der Meer and Ernest Lawrence are the only two accelerator physicists awarded with the Nobel Prize.

STOCHASTIC COOLING (1984)

Stochastic cooling is a form of particle beam cooling. It is used in some particle accelerators and storage rings to control the emittance of the particle beams in the machine. This process uses the electrical signals that the individual charged particles generate in a feedback loop to reduce the tendency of individual particles to move away from the other particles in the beam. It is accurate to think of this as thermodynamic cooling, or the reduction of entropy, in much the same way that a refrigerator or an air conditioner cools its contents.

The technique was invented and applied at the Intersecting Storage Rings, and later the Super Proton Synchrotron, at CERN in Geneva, Switzerland by Simon van der Meer, an engineer from the Netherlands. It was used to collect and cool antiprotons–these particles were injected into the SPS with counter-rotating protons and collided at a particle physics experiment. For this work, van der Meer was awarded the Nobel Prize in Physics in 1984. He shared this prize with Carlo Rubbia of Italy, who conducted the physics experiment that took advantage of this breakthrough. This experiment discovered the W and Z bosons, fundamental particles that carry the weak nuclear force.

Fermi National Accelerator Laboratory continues to use *stochastic cooling* in its antiproton source. The accumulated antiprotons are used in the Tevatron to collide with protons to create collisions at CDF and the D0 (D zero) experiment.

Stochastic cooling in the Tevatron at Fermi lab was attempted, but was not fully successful. The equipment was subsequently sold to Brookhaven National Laboratory, where it was successfully employed in 2007, in the RHIC.

Klaus von Klitzing

 Klaus von Klitzing is a German physicist known for discovery of the integer quantum Hall Effect, for which he was awarded the 1985 Nobel Prize in Physics. In 1962, von Klitzing passed the Abitur at Artland Gymnasium in Quakenbrück, Germany, before studying physics at the Technical University of Braunschweig, where he received his diploma in 1969.

He continued his studies at the University of Würzburg, completing his PhD thesis Galvanomagnetic Properties of Tellurium in Strong Magnetic Fields in 1972, and habilitation in 1978. This work was performed at the Clarendon Laboratory in Oxford and the Grenoble High Magnetic Field Laboratory in France, where he continued to work until becoming a professor at the Technical University of Munich in 1980. Von Klitzing has been a director of the Max Planck Institute for Solid State Research in Stuttgart since 1985.

Born	28 June 1943 (age 66), Schroda (Posen)
Nationality	Germany
Fields	Physics
Known for	Quantum hall effect
Notable awards	Nobel Prize in Physics (1985)

QUANTUM HALL EFFECT (1985)

The quantum Hall effect (or integer quantum Hall effect) is a quantum-mechanical version of the Hall effect, observed in two-dimensional electron systems subjected to low temperatures and strong magnetic fields, in which the Hall conductivity σ takes on the quantized values $\sigma = v\dfrac{e^2}{h}$, where e is the elementary charge and h is Planck's constant.

The prefactor v is known as the filling factor, and can take on either integer ($v = 1, 2, 3, ..$) or rational fraction ($v = 1/3, 1/5, 5/2, 12/5 ..$) values. The quantum Hall effect is referred to as the integer or fractional quantum Hall effect depending on whether v is an integer or fraction respectively. The integer quantum Hall effect is very well understood, and can be simply explained in terms of single particle orbitals of an electron in a magnetic field fractional quantum Hall effect, however, is more complicated, and its existence relies fundamentally on electron-electron interactions.

The integer quantization of the Hall conductance was originally predicted by Ando, Matsumoto, and Uemura in 1975, on the basis of an approximate calculation which they themselves did not believe to be true. Several workers subsequently observed the effect in experiments carried out on the inversion layer of MOSFETs. It was only in 1980 that Klaus von Klitzing, working with samples developed by Michael Pepper and Gerhard Dorda, made the unexpected discovery that the Hall conductivity was *exactly* quantized. For this finding, von Klitzing was awarded the 1985 Nobel Prize in Physics. The link between exact quantization and gauge invariance was subsequently found by Robert Laughlin. Most integer quantum Hall experiments are now performed on gallium arsenide heterostructures, although many other semiconductor materials can be used. The integer quantum Hall effect has also been found in graphene at temperatures as high as room temperature.

Ernst August Friedrich Ruska

Ernst August Friedrich Ruska was a German physicist who won the Nobel Prize in Physics in 1986 for his work in electron optics, including the design of the first electron microscope.

Born	25 December 1906, Heidelberg, Germany
Died	27 May 1988 (age 81), West Berlin, Germany
Nationality	Germany
Fields	Physics
Institutions	Fritz Haber Institute, Technical University of Berlin
Alma mater	Technical University of Munich
Doctoral advisor	Max Knoll
Known for	Electron Microscopy
Notable awards	Nobel Prize in Physics (1986)

ELECTRON MICROSCOPE (1986)

In 1931, the German physicist Ernst Ruska and German electrical engineer Max Knoll constructed the prototype electron microscope, capable of four-hundred-power magnification; the apparatus was a practical application of the principles of electron microscopy. Two years later, in 1933, Ruska built an electron microscope that exceeded the resolution attainable with an optical (lens) microscope. Moreover, Reinhold Rudenberg, the scientific director of Siemens-Schuckertwerke, obtained the patent for the electron microscope in May of 1931. Family illness compelled the electrical engineer to devise an electrostatic microscope, because he wanted to make visible the poliomyelitis virus.

In 1937, the Siemens company financed the development work of Ernst Ruska and Bodo von Borries, and employed Helmut Ruska (Ernst's brother) to develop applications for the microscope, especially with biologic specimens. Also in 1937, Manfred von Ardenne pioneered the scanning electron microscope. The first *practical* electron microscope was constructed in 1938, at the University of Toronto, by Eli Franklin Burton and students Cecil Hall, James Hillier, and Albert Prebus; and Siemens produced the first *commercial* Transmission Electron Microscope (TEM) in 1939. Although contemporary electron microscopes are capable of two million-power magnification, as scientific instruments, they remain based upon Ruska's prototype.

An electron microscope is a type of microscope that produces an electronically-magnified image of a specimen for detailed observation. The electron microscope (EM) uses a particle beam of electrons to illuminate the specimen and create a magnified image of it. The microscope has a greater resolving power than a light-powered optical microscope, because it uses electrons that have wavelengths about 100,000 times shorter than visible light (photons), and can achieve magnifications of up to 1,000,000 ×, whereas light microscopes are limited to 2000 × magnification.

The electron microscope uses electrostatic and electro-magnetic lenses to control the electron beam and focus it to form an image. These lenses are analogous to, but different from the glass lenses of an optical microscope that form a magnified image by focusing light on or through the specimen.

Electron microscopes are used to observe a wide range of biological and inorganic specimens including microorganisms, cells, large molecules, biopsy samples, metals, and crystals. Industrially, the electron microscope is primarily used for quality control and failure analysis in semiconductor device fabrication.

Gerd Binnig

Gerd Binnig is a German physicist, and a Nobel laureate. He was born in Frankfurt am Main and played in the ruins of the city during his childhood. His family lived partly in Frankfurt and partly in Offenbach am Main, and he attended school in both the cities.

At the age of 10, he decided to become a physicist, but he soon wondered whether he had made the right choice. He concentrated more on music, playing in a band. He also started playing the violin at 15 and played in his school orchestra. The team included Christoph Gerber and Edmund Weibel, and they were soon recognized with a number of prizes: the German Physics Prize, the Otto Klung Prize, the Hewlett Packard Prize, the King Faisal Prize, and ultimately, the Nobel Prize.

Born	20 July 1947, Frankfurt am Main
Fields	Physics
Known for	Scanning tunneling microscope
Notable awards	Nobel Prize in Physics (1986) The Elliott Cresson Medal (1987)

SCANNING TUNNELING MICROSCOPE (1986)

A scanning tunneling microscope (STM) is a powerful instrument for imaging surfaces at the atomic level. Its development in 1981 earned its inventors, Gerd Binnig and Heinrich Rohrer (at IBM Zürich), the Nobel Prize in Physics in 1986. For an STM, good resolution is considered to be 0.1 nm lateral resolution and 0.01 nm depth resolution. With this resolution, individual atoms within materials are routinely imaged and manipulated. The STM can be used not only in ultra high vacuum but also in air, water, and various other liquid or gas ambients, and at temperatures ranging from near zero Kelvin to a few hundred degrees Celsius.

The STM is based on the concept of quantum tunneling. When a conducting tip is brought very near to the surface to be examined, a bias (voltage difference) applied between the two can allow electrons to tunnel through the vacuum between them. The resulting *tunneling current* is a function of tip position, applied voltage, and the local density of states (LDOS) of the sample. Information is acquired by monitoring the current as the tip's position scans across the surface, and is usually displayed in image form. STM can be a challenging technique, as it requires extremely clean and stable surfaces, sharp tips, excellent vibration control, and sophisticated electronics.

First, a voltage bias is applied and the tip is brought close to the sample by some coarse sample-to-tip control, which is turned off when the tip and sample are sufficiently close. At close range, fine control of the tip in all three dimensions when near the sample is typically piezoelectric, maintaining tip-sample separation W typically in the 4-7 Å range, which is the equilibrium position between attractive ($3<W<10$Å) and repulsive ($W<3$Å) interactions. In this situation, the voltage bias will cause electrons to tunnel between the tip and sample, creating a current that can be measured. Once tunneling is established, the tip's bias and position with respect to the sample can be varied (with the details of this variation depending on the experiment) and data is obtained from the resulting changes in current.

If the tip is moved across the sample in the x-y plane, the changes in surface height and density of states cause changes in current. These changes are mapped in images. This change in current with respect to position can be measured itself, or the height, z, of the tip corresponding to a constant current can be measured. These two modes are called constant height mode and constant current mode, respectively. In constant current mode, feedback electronics adjust the height by a voltage to the piezoelectric height control mechanism. This leads to a height variation and thus the image comes from the tip topography across the sample and gives a constant charge density surface; this means contrast on the image is due to variations in charge density.

Heinrich Rohrer

Heinrich Rohrer is a Swiss physicist who shared the 1986 Nobel Prize in Physics with Gerd Binnig for the design of the scanning tunneling microscope (STM).

Born	6 June 1933 (age 76), St. Gallen
Nationality	Swiss
Fields	Physics
Known for	Scanning tunneling microscope
Notable awards	Nobel Prize in Physics (1986), Elliott Cresson Medal (1987)

SCANNING TUNNELING MICROSCOPE (1986)

The components of an STM include scanning tip, piezoelectric controlled height and x,y scanner, coarse sample to tip control, vibration isolation system, and computer.

The resolution of an image is limited by the radius of curvature of the scanning tip of the STM. Additionally, image artifacts can occur if the tip has two tips at the end rather than a single atom; this leads to double-tip imaging, a situation in which both tips contribute to the tunneling. Therefore it has been essential to develop processes for consistently obtaining sharp, usable tips. Recently, carbon nanotubes have been used in this instance.

The tip is often made of tungsten or platinum-iridium, though gold is also used. Tungsten tips are usually made by electrochemical etching, and platinum-iridium tips by mechanical shearing.

Due to the extreme sensitivity of tunnel current to height, proper vibration isolation or an extremely rigid STM body is imperative for obtaining usable results. In the first STM by Binnig and Rohrer, magnetic levitation was used to keep the STM free from vibrations; now mechanical spring or gas spring systems are often used. Additionally, mechanisms for reducing eddy currents are sometimes implemented.

Maintaining the tip position with respect to the sample, scanning the sample and acquiring the data is computer controlled. The computer may also be used for enhancing the image with the help of image processing as well as performing quantitative measurements.

Many other microscopy techniques have been developed based upon STM. These include photon scanning microscopy (PSTM), which uses an optical tip to tunnel photons; scanning tunneling potentiometry (STP), which measures electric potential across a surface; spin polarized scanning tunneling microscopy (SPSTM), which uses a ferromagnetic tip to tunnel spin-polarized electrons into a magnetic sample, and atomic force microscopy (AFM), in which the force caused by interaction between the tip and sample is measured.

Other STM methods involve manipulating the tip in order to change the topography of the sample. This is attractive for several reasons. Firstly the STM has an atomically precise positioning system which allows very accurate atomic scale manipulation. Furthermore, after the surface is modified by the tip, it is a simple matter to then image with the same tip, without changing the instrument. IBM researchers developed a way to manipulate Xenon atoms adsorbed on a nickel surface This technique has been used to create electron corrals with a small number of adsorbed atoms, which allows the STM to be used to observe electron Friedel oscillations on the surface of the material. Aside from modifying the actual sample surface, one can also use the STM to tunnel electrons into a layer of E-Beam photo resist on a sample, in order to do lithography. This has the advantage of offering more control of the exposure than traditional Electron beam lithography. Another practical application of STM is atomic deposition of metals (Au, Ag, W, etc.) with any desired (pre-programmed) pattern, which can be used as contacts to nanodevices or as nanodevices themselves.

Johannes Georg Bednorz

 Johannes Georg Bednorz is a physicist at the IBM Zürich Research Laboratory. He is best known for his role in the discovery of high-temperature superconductivity, for which he shared the 1987 Nobel Prize in Physics.

Born	16 May 1950 (age 58), Neuenkirchen, North Rhine-Westphalia, Germany
Nationality	German
Fields	Physics
Doctoral advisor	Heini Gränicher, Karl Alexander Müller
Known for	High-temperature superconductivity
Notable awards	Nobel Prize in Physics (1987)

HIGH-TEMPERATURE SUPERCONDUCTIVITY (1987)

High-temperature superconductors are materials that have a superconducting transition temperature (T_c) above 30 K. From 1960 to 1980, 30 K was thought to be the highest theoretically possible T_c. The first high-T_c superconductor was discovered in 1986 by IBM Researchers Karl Müller and Johannes Bednorz, for which they were awarded the Nobel Prize in Physics in 1987. Until Fe-based superconductors were discovered in 2008, the term high-temperature superconductor was used interchangeably with cuprate superconductor for compounds such as bismuth strontium calcium copper oxide and yttrium barium copper oxide.

High-temperature has three common definitions in the context of superconductivity:

1. Above the temperature of 30 K that had historically been taken as the upper limit allowed by BCS theory. This is also above the 1973 record of 23 K that had lasted until copper-oxide materials were discovered in 1986.

2. Having a transition temperature that is a larger fraction of the Fermi temperature than for conventional super-conductors such as elemental mercury or lead. This definition encompasses a wider variety of unconventional superconductors and is used in the context of theoretical models.

3. Greater than the boiling point of liquid nitrogen (77 K or −196 °C). This is significant for technological applications of superconductivity because liquid nitrogen is a relatively inexpensive and easily handled coolant.

Technological applications benefit from both the higher critical temperature being above the boiling point of liquid nitrogen and also the higher critical magnetic field at which superconductivity is destroyed. In magnet applications the high critical magnetic field may be more valuable than the high T_c itself. Some cuprates have an upper critical field around 100 teslas. However, cuprate materials are brittle ceramics which are expensive to manufacture and not easily turned into wires or other useful shapes.

Two decades of intense experimental and theoretical research, with over 100,000 published papers on the subject, have discovered many common features in the properties of high-temperature superconductors, but as of 2009, there is no widely accepted theory to explain their properties. Cuprate superconductors differ in many important ways from conventional superconductors, such as elemental mercury or lead, which are adequately explained by the BCS theory. There also has been much debate as to high-temperature superconductivity coexisting with magnetic ordering in YBCO, iron-based superconductors, several ruthenocuprates and other exotic superconductors, and the search continues for other families of materials. HTS are Type-II superconductors, which allow magnetic fields to penetrate their interior in quantized units of flux, meaning that much higher magnetic fields are required to suppress superconductivity. The layered structure also gives a directional dependence to the magnetic field response.

Karl Alexander Müller

 Karl Alexander Müller is a Swiss physicist and Nobel laureate. He received the Nobel Prize in Physics in 1987 with Johannes Georg Bednorz for their work in superconductivity in ceramic materials.

Born	20 April 1927 (age 83), Basel, Switzerland
Nationality	Swiss
Fields	Physics
Institutions	IBM Zürich Research Laboratory, University of Zurich Battelle Memorial Institute
Alma mater	ETH Zürich
Known for	High-temperature superconductivity
Notable awards	Nobel Prize in Physics (1987)

Cuprate superconductors are generally considered to be quasi-two-dimensional materials with their superconducting properties determined by electrons moving within weakly coupled copper-oxide (CuO_2) layers. Neighbouring layers containing ions such as lanthanum, barium, strontium, or other atoms act to stabilize the structure and dope electrons or holes onto the copper-oxide layers. The undoped 'parent' or 'mother' compounds are Mott insulators with long-range antiferromagnetic order at low enough temperature. Single band models are generally considered to be sufficient to describe the electronic properties.

The cuprate superconductors adopt a perovskite structure. The copper-oxide planes are checkerboard lattices with squares of O^{2-} ions with a Cu^{2+} ion at the centre of each square. The unit cell is rotated by $45°$ from these squares. Chemical formulae of superconducting materials generally contain fractional numbers to describe the doping required for superconductivity. There are several families of cuprate superconductors and they can be categorized by the elements they contain and the number of adjacent copper-oxide layers in each superconducting block. For example, YBCO and BSCCO can alternatively be referred to as Y123 and Bi2201/Bi2212/Bi2223 depending on the number of layers in each superconducting block (n). The superconducting transition temperature has been found to peak at an optimal doping value (p = 0.16) and an optimal number of layers in each superconducting block, typically n = 3.

HISTORY AND PROGRESS (1987)

- April 1986–The term *high-temperature superconductor* was first used to designate the new family of cuprate-perovskite ceramic materials discovered by Johannes Georg Bednorz and Karl Alexander Müller, for which they won the Nobel Prize in Physics the following year. Their discovery of the first high-temperature superconductor, LaBaCuO, with a transition temperature of 30 K, generated great excitement.
- LSCO ($La_{2-x}Sr_xCuO_4$) discovered the same year.
- January 1987: YBCO was discovered to have a T_c of 90 K.
- 1988: BSCCO (Bismuth Strontium Calcium Copper Oxide) discovered with T_c up to 108 K, and TBCCO (T = thallium) discovered to have T_c of 127 K.
- As of 2009, the highest-temperature superconductor (at ambient pressure) is mercury barium calcium copper oxide ($HgBa_2Ca_2Cu_3O_x$), at 135 K and is held by a cuprate-perovskite material, possibly 164 K under high pressure.
- Recently, iron-based superconductors with critical temperatures as high as 56 K have been discovered. These are often also referred to as high-temperature superconductors.

After more than twenty years of intensive research the origin of high-temperature superconductivity is still not clear, but it seems that instead of *electron-phonon* attraction mechanisms, as in conventional superconductivity, one is dealing with genuine *electronic* mechanisms (e.g. by antiferromagnetic correlations), and instead of s-wave pairing, d-waves are substantial. One goal of all this research is room-temperature superconductivity.

Leon Max Lederman

Leon Max Lederman is an American experimental physicist and Nobel Prize in Physics laureate for his work with neutrinos. In 1988, Lederman received the Nobel Prize for Physics along with Melvin Schwartz and Jack Steinberger for the neutrino beam method and the demonstration of the doublet structure of the leptons through the discovery of the muon neutrino.

Born	15 July 1922 (age 87), New York
Nationality	United States
Fields	Physics
Known for	Neutrinos, bottom quark
Notable awards	Nobel Prize in Physics (1988), National Medal of Science (1965), The Elliott Cresson Medal for Physics (1976), The Wolf Prize for Physics (1982) and The Enrico Fermi Award (1992).

NEUTRINO (1988)

A neutrino (meaning small neutral one) is an elementary particle that usually travels close to the speed of light, is electrically neutral, and is able to pass through ordinary matter almost undisturbed. This makes neutrinos extremely difficult to detect. Neutrinos have a very small, but nonzero mass. They are denoted by the Greek letter ν (nu).

Neutrinos are created as a result of certain types of radioactive decay or nuclear reactions such as those that take place in the Sun, in nuclear reactors, or when cosmic rays hit atoms. There are three types, or flavors, of neutrinos: electron neutrinos, muon neutrinos and tau neutrinos; each type also has a corresponding antiparticle, called antineutrinos. Electron neutrinos (or antineutrinos) are generated whenever protons change into neutrons (or vice versa), the two forms of beta decay. Interactions involving neutrinos are mediated by the weak interaction.

Most neutrinos passing through the Earth emanate from the Sun, and more than 50 trillion solar neutrinos pass through the human body every second.

BOTTOM QUARK

The bottom quark, also known as the beauty quark, is a third-generation quark with a charge of $-\frac{1}{3}$ e. Although all quarks are described in a similar way by the quantum chromo dynamics, the bottom quark's large mass (around 4,200 Mev/c^2, a bit more than four times the mass of a proton), combined with low values of the CKM matrix elements V_{ub} and V_{cb}, gives it a distinctive signature that makes it relatively easy to identify experimentally (using a technique called B-tagging). Because three generations of quark are required for CP-violation mesons containing the bottom quark are the easiest particles to use to investigate the phenomenon; such experiments are being performed at the BaBar and Belle experiments. The bottom quark is also notable because it is a product in almost all top quark decays, and would be a frequent decay product for the hypothetical Higgs boson if it is sufficiently light.

The bottom quark was theorized in 1973 by physicists Makoto Kobayashi and Toshihide Maskawa to explain CP-violation. The name bottom was introduced in 1975 by Haim Harari. The bottom quark discovered in 1977 by the Fermi lab E288 experiment team led by Leon M. Lederman, when collisions produced bottomonium. Kobayashi and Maskawa won the 2008 Nobel Prize in Physics for their explanation of CP-violation. On its discovery, there were efforts to name the bottom quark beauty, but bottom became the predominant usage.

Melvin Schwartz

Melvin Schwartz was an American physicist. He shared the 1988 Nobel Prize in Physics with Leon M. Lederman and Jack Steinberger for their development of the neutrino beam method and their demonstration of the doublet structure of the leptons through the discovery of the muon neutrino.

Born	2 November 1932, New York City, New York
Died	28 August 2006 (age 73), Twin Falls, Idaho
Nationality	American
Fields	Particle physics
Institutions	Brookhaven National Laboratory, Stanford University, Columbia University
Alma mater	Columbia University
Known for	Neutrinos
Notable awards	Nobel Prize in Physics (1988)

EXPERIMENTAL DEMONSTRATION OF NEUTRINO FLAVORS (1988)

In 1962 Leon M. Lederman, Melvin Schwartz and Jack Steinberger showed that more than one type of neutrino exists by first detecting interactions of the muon neutrino (already hypothesized with the name *neutretto*, which earned them the 1988 Nobel Prize. When the third type of lepton, the tau, was discovered in 1975 at the Stanford Linear Accelerator Center, it too was expected to have an associated neutrino (the tau neutrino). First evidence for this third neutrino type came from the observation of missing energy and momentum in tau decays analogous to the beta decay leading to the discovery of the neutrino. The first detection of tau neutrino interactions was announced in summer of 2000 by the DONUT collaboration at Fermi lab, making it the latest particle of the Standard Model to have been directly observed; its existence had already been inferred by both theoretical consistency and experimental data from the Large Electron-Positron Collider.

In 1942 Kan-Chang Wang first proposed the use of beta-capture to experimentally detect neutrinos. In 1956 Clyde Cowan, Frederick Reines, F. B. Harrison, H. W. Kruse, and A. D. McGuire detected the neutrino through this process, a result that was rewarded with the 1995 Nobel Prize. In this experiment, now known as the Cowan-Reines neutrino experiment, neutrinos created in a nuclear reactor by beta decay were shot into protons producing neutrons and positrons both of which could be detected. It is now known that both the proposed and the observed particles were antineutrinos.

Starting in the late 1960s, several experiments found that the number of electron neutrinos arriving from the Sun was between one third and one half the number predicted by the Standard Solar Model. This discrepancy, which became to be known as the solar neutrino problem, remained unresolved for some thirty years. The Standard Model of particle physics assumes that neutrinos are massless and cannot change flavor. However, if neutrinos had mass, they could change flavour (or oscillate between flavours).

A practical method for investigating neutrino oscillations was first suggested by Bruno Pontecorvo in 1957 using an analogy with kaon oscillations; over the subsequent 10 years he developed the mathematical formalism and the modern formulation of vacuum oscillations. In 1985 Stanislav Mikheyev and Alexei Smirnov (expanding on 1978 work by Lincoln Wolfenstein) noted that flavour oscillations can be modified when neutrinos propagate through matter. This so-called Mikheyev-Smirnov-Wolfenstein effect (MSW effect) is important to understand because many neutrinos emitted by fusion in the Sun pass through the dense matter in the solar core (where essentially all solar fusion takes place) on their way to detectors on Earth.

Jack Steinberger

 Jack Steinberger is a German-American physicist currently residing near Geneva, Switzerland. He co-discovered the muon neutrino, along with Leon Lederman and Melvin Schwartz, for which they were given the 1988 Nobel Prize in Physics.

Born	25 May 1921 (age 89), Bad Kissingen
Nationality	Germany, United States, Switzerland
Fields	Physics
Known for	Discovery of the muon neutrino
Notable awards	Nobel Prize in Physics (1988)

MUON NEUTRINO (1988)

The muon neutrino (v_μ) is the second of the three neutrinos. It, along with the muon, forms the second generation of leptons, hence its name muon neutrino. It was first hypothesized by in the early 1940s by several people, and was discovered in 1962 by Leon Lederman, Melvin Schwartz and Jack Steinberger. The discovery was rewarded with the 1988 Nobel Prize in Physics.

In 1962 Leon M. Lederman, Melvin Schwartz and Jack Steinberger showed that more than one type of neutrino exists by first detecting interactions of the muon neutrino (already hypothesized with the name neutretto), which earned them the 1988 Nobel Prize.

During 1954–1955, Steinberger contributed to the development of the bubble chamber with the construction of a 15 cm device for use with the Cosmotron at Brookhaven. The experiment used a pion beam to produce pairs of hadrons with strange quarks in order to elucidate the puzzling production and decay properties of these particles. In 1956, a 30 cm chamber outfitted with three cameras was used in the discovery of the neutral Sigma hyperon and a measurement of its mass. This observation was important for confirming the existence of the SU(3) flavor symmetry which hypothesizes the existence of the strange quark.

An important characteristic of the weak interaction is its violation of parity symmetry. This characteristic was established through the measurement of the spins and parities of many hyperons. Steinberger and his collaborators contributed several such measurements using large (75 cm) liquid-hydrogen bubble chambers and separated hadron beams at Brookhaven. One example is the measurement of the invariant mass distribution of electron-positron pairs produced in the decay of Sigma-zero hyperons to Lambda-zero hyperons.

The neutrino has half-integer spin ($\frac{1}{2}\hbar$) and is therefore a fermion. Neutrinos interact primarily through the weak force. The discovery of neutrino flavor oscillations implies that neutrinos have mass. The existence of a neutrino mass strongly suggests the existence of a tiny neutrino magnetic moment of the order of $10^{-19}\mu B$, allowing the possibility that neutrinos may interact electromagnetically as well. An experiment done by C. S. Wu at Columbia University showed that neutrinos always have left-handed chirality.

It is very hard to uniquely identify neutrino interactions among the natural background of radioactivity. For this reason, in early experiments a special reaction channel was chosen to facilitate the identification: the interaction of an antineutrino with one of the hydrogen nuclei in the water molecules. A hydrogen nucleus is a single proton, so simultaneous nuclear interactions, which would occur within a heavier nucleus; do not need to be considered for the detection experiment. Within a cubic metre of water placed right outside a nuclear reactor, only relatively few such interactions can be recorded, but the setup is now used for measuring the reactor's plutonium production rate.

Norman Foster Ramsey

 Norman Foster Ramsey, Jr. is an American physicist. A physics professor at Harvard University since 1947, Ramsey also held several posts with such government and international agencies as NATO and the United States Atomic Energy Commission.

He was awarded the 1989 Nobel Prize in Physics for the invention of the separated oscillatory field method, which had important applications in the construction of atomic clocks. Ramsey shared the prize with Hans G. Dehmelt and Wolfgang Paul.

Born	27 August 1915 (age 94), Washington, DC
Residence	United States
Nationality	United States
Fields	Physics
Institutions	Harvard University
Alma mater	Columbia University, University of Cambridge
Known for	Separated oscillatory field method
Notable awards	IEEE Medal of Honor, Nobel Prize in Physics (1989)

ATOMIC CLOCK (1989)

An atomic clock is a type of clock that uses an atomic resonance frequency standard as its timekeeping element. They are the most accurate time and frequency standards known, and are used as primary standards for international time distribution services, to control the frequency of television broadcasts, and in global navigation satellite systems such as GPS.

Atomic clocks do not use radioactivity, but rather the precise microwave signal that electrons in atoms emit when they change energy levels. Early atomic clocks were based on masers. Currently, the most accurate atomic clocks are based on absorption spectroscopy of cold atoms in atomic fountains such as the NIST-F1.

National standards agencies maintain an accuracy of 10^{-9} seconds per day (approximately 1 part in 10^{14}), and a precision set by the radiotransmitter pumping the maser. The clocks maintain a continuous and stable time scale, International Atomic Time (TAI). For civil time, another time scale is disseminated, Coordinated Universal Time (UTC). UTC is derived from TAI, but synchronized, by using leap seconds, to UT1, which is based on actual rotations of the earth with respect to the solar time.

The idea of using atomic transitions to measure time was first suggested by Lord Kelvin in 1879. The practical method for doing this became magnetic resonance, developed in the 1930s by Isidor Rabi. In 1945, Rabi first publicly suggested that atomic beam magnetic resonance might be used as the basis of a clock. The first atomic clock was an ammonia maser device built in 1949 at the US National Bureau of Standards (NBS, now NIST). It was less accurate than existing quartz clocks, but served to demonstrate the concept. The first accurate atomic clock, a caesium standard based on a certain transition of the caesium-133 atom, was built by Louis Essen in 1955 at the National Physical Laboratory in the UK. Calibration of the caesium standard atomic clock was carried out by the use of the astronomical time scale ephemeris time (ET). This led to the internationally agreed definition of the latest SI second being based on atomic time. Equality of the ET second with the (atomic clock) SI second has been verified to within 1 part in 10^{10}. The SI second thus inherits the effect of decisions by the original designers of the ephemeris time scale, determining the length of the ET second.

Hans Georg Dehmelt

 Hans Georg Dehmelt is a German born American physicist, who co-developed the ion trap technique with Wolfgang Paul, for which they both received the Nobel Prize in Physics in 1989. The technique was used for high precision measurement of the electron g-factor.

Born	9 September 1922 (age 87), Görlitz, Germany
Residence	United States
Nationality	Germany
Fields	Physics
Institutions	Duke University, University of Washington
Alma mater	University of Göttingen, Duke University
Known for	Development of the ion trap, precise measurement of the electron g-factor
Notable awards	Nobel Prize in Physics (1989)

ION TRAP (1989)

An ion trap is a combination of electric or magnetic fields that captures ions in a region of a vacuum system or tube. Ion traps have a number of scientific uses such as mass spectrometry and trapping ions while the ion's quantum state is manipulated. The two most common types of ion traps are the Penning trap and the Paul trap (quadrupole ion trap).

When using ion traps for scientific studies of quantum state manipulation, the Paul trap is most often used. This work may lead to a trapped ion quantum computer and has already been used to create the world's most accurate atomic clocks.

An ion trap mass spectrometer may incorporate a Penning trap (Fourier transform ion cyclotron resonance), Paul trap or the Kingdon trap. The Orbitrap, introduced in 2005, is based on the Kingdon trap. Other types of mass spectrometers may also use a linear quadrupole ion trap as a selective mass filter.

In an electron gun (a device emitting high-speed electrons, such as those in CRTs), an ion trap may be implemented above the cathode (using an extra, positively-charged electrode between the cathode and the extraction electrode) to prevent its degradation by positive ions accelerated backward by the fields intended to pull electrons away from the cathode.

A g-factor (also called g value or dimensionless magnetic moment) is a dimensionless quantity which characterizes the magnetic moment and gyromagnetic ratio of a particle or nucleus. It is essentially a proportionality constant that relates the observed magnetic moment μ of a particle to the appropriate angular momentum quantum number and the appropriate fundamental quantum unit of magnetism, usually the Bohr magneton or nuclear magneton.

There are three magnetic moments associated with an electron: One from its spin angular momentum, one from its orbital angular momentum, and one from its total angular momentum (the quantum-mechanical sum of those two components). Corresponding to these three moments are three different g-factors: Electron spin g-factor, Electron orbital g-factor, Landé g-factor.

The g factor, where g stands for general intelligence, is a statistic used in psychometrics in an attempt to quantify the mental ability underlying results of various tests of cognitive ability. The existence of such an underlying g factor was postulated in 1904 by Charles Spearman.

Spearman, who was an early psychometrician, found that school children's grades across seemingly unrelated subjects were positively correlated, and proposed that these correlations reflected the influence of a dominant factor, which he termed g for general intelligence or ability. He developed a model in which all variations in intelligence test scores are explained by two factors: first, a factor specific to an individual mental task: the individual abilities that would make a person more skilled at a specific cognitive task; and second a general factor g that governs performance on all cognitive tasks.

Wolfgang Paul

Wolfgang Paul was a German physicist, who co-developed the ion trap. He received the Nobel Prize in Physics in 1989 for this work.

Born	10 August 1913, Lorenzkirch, Saxony, Germany
Died	7 December 1993 (age 80), Bonn, North Rhine-Westphalia, Germany
Nationality	Germany
Fields	Physics
Institutions	University of Bonn
Alma mater	Technical University of Munich, Technical University of Berlin, University of Göttingen
Doctoral advisor	Hans Kopfermann
Known for	Ion traps
Notable awards	Nobel Prize in Physics (1989)

Electron g-factors: There are three magnetic moments associated with an electron: one from its spin angular momentum, one from its orbital angular momentum, and one from its total angular momentum (the quantum-mechanical sum of those two components). Corresponding to these three moments are three different g-factors:

Electron spin g-factor: The most famous of these is the electron spin g-factor (more often called simply the electron g-factor), g_e, defined by $\mu_s = g_e \mu_B S/\hbar$

where μ_s is the total magnetic moment resulting from the spin of an electron, S is the magnitude of its spin angular momentum, and μ_B is the Bohr magneton. In atomic physics, the electron spin g-factor is often defined as the absolute value or negative of g_e:

$g_S = |g_e| = -g_e$.

The z-component of the magnetic moment then becomes $\mu z = -g_S \mu_B m_s$.

The value g_S is roughly equal to 2.002319, and is known to extraordinary accuracy. The reason it is not precisely two is explained by quantum electrodynamics calculation of the anomalous magnetic dipole moment.

Electron orbital g-factor: Secondly, the electron orbital g-factor, g_L, is defined by $\mu_L = g_L \mu_B L/\hbar$, where μ_L is the total magnetic moment resulting from the orbital angular momentum of an electron, L is the magnitude of its orbital angular momentum, and μ_B is the Bohr magneton. The value of g_L is exactly equal to one, by a quantum-mechanical argument analogous to the derivation of the classical magnetogyric ratio. For an electron in an orbital with a magnetic quantum number m_l, the z-component of the orbital angular momentum is $\mu z = g_L \mu_B m_l$ which, since $g_L = 1$, is just $\mu_B m_l$.

Landé g-factor: Thirdly, the Landé g-factor, g_J, is defined by $\mu = g_J \mu_B J/\hbar$, where μ is the total magnetic moment resulting from both spin and orbital angular momentum of an electron, $J = L + S$ is its total angular momentum, and μ_B is the Bohr magneton. The value of g_J is related to g_L and g_S by a quantum-mechanical argument.

There is some debate about whether g represents a real thing or is just a sort of statistical average of test results. The accumulation of cognitive testing data and improvements in analytical techniques have preserved g's central role but have led to a modern conception of g. According to the American Psychological Association, a hierarchy of factors with g at its apex and group factors at successively lower levels is now the most widely accepted model of cognitive ability. Other models have also been proposed, and significant controversy attends g and its alternatives.

Jerome Isaac Friedman

Jerome Isaac Friedman is an American physicist. He was born in Chicago, Illinois to parents who emigrated to the US from Russia, and excelled particularly in art while growing up.

He became interested in physics after reading a book on relativity written by Albert Einstein, and as a result he turned down a scholarship to the Art Institute of Chicago to study physics at the University of Chicago. While there he worked under Enrico Fermi, and eventually received his PhD in physics in 1956. In 1960 he joined the physics faculty of the Massachusetts Institute of Technology.

Born	28 March 1930 (age 80), Chicago, Illinois
Nationality	United States
Fields	Physics
Institutions	MIT
Alma mater	Chicago
Doctoral advisor	Enrico Fermi
Known for	Experimental proof of quarks
Notable awards	Nobel Prize in Physics (1990)

QUARK (1990)

A quark is an elementary particle and a fundamental constituent of matter. Quarks combine to form composite particles called hadrons, the most stable of which are protons and neutrons, the components of atomic nuclei. Due to a phenomenon known as colour confinement, quarks are never found in isolation; they can only be found within hadrons. For this reason, much of what is known about quarks has been drawn from observations of the hadrons themselves.

There are six types of quarks, known as flavors: up, down, charm, strange, top, and bottom. Up and down quarks have the lowest masses of all quarks. The heavier quarks rapidly change into up and down quarks through a process of particle decay: the transformation from a higher mass state to a lower mass state. Because of this, up and down quarks are generally stable and the most common in the universe, whereas charm, strange, top, and bottom quarks can only be produced in high energy collisions (such as those involving cosmic rays and in particle accelerators).

Quarks have various intrinsic properties, including electric charge, colour charge, spin, and mass. Quarks are the only elementary particles in the Standard Model of particle physics to experience all four fundamental interactions, also known as fundamental forces (electromagnetism, gravitation, strong interaction, and weak interaction), as well as the only known particles whose electric charges are not integer multiples of the elementary charge. For every quark flavour there is a corresponding type of antiparticle, known as antiquark, that differs from the quark only in that some of its properties have equal magnitude but opposite sign.

The quark model was independently proposed by physicists Murray Gell-Mann and George Zweig in 1964. Quarks were introduced as parts of an ordering scheme for hadrons, and there was little evidence for their physical existence until 1968. All six flavors of quark have since been observed in accelerator experiments; the top quark, first observed at Fermi lab in 1995, was the last to be discovered.

Henry Way Kendall

Henry Way Kendall was an American particle physicist who won the Nobel Prize in Physics in 1990 jointly with Jerome Isaac Friedman and Richard E. Taylor for their pioneering investigations concerning deep inelastic scattering of electrons on protons and bound neutrons, which have been of essential importance for the development of the quark model in particle physics.

Born	9 December 1926, Boston, Massachusetts
Died	15 February 1999 (age 72), Wakulla Springs State Park, Florida
Nationality	United States
Fields	Physics
Institutions	MIT
Alma mater	Amherst College, MIT
Doctoral advisor	Martin Deutsch
Notable awards	Nobel Prize in Physics (1990)

QUARK MODEL (1990)

In physics, the quark model, originally just a very good classification scheme to organize the depressingly large number of hadrons that were being discovered starting in the 1950s and continuing through the 1960s, received experimental verification beginning in the late 1960s and continuing to the present. Hadrons are not fundamental, but their valence quarks are thought to be, the quarks and antiquarks which give rise to the quantum numbers of the hadrons. These quantum numbers are labels identifying the hadrons, and are of two kinds. One set comes from the Poincaré symmetry, J^{PC}, where J, P and C stand for the total angular momentum, P-symmetry, and C-symmetry respectively. The remainder are flavour quantum numbers such as the isospin, strangeness, charm, and so on. The quark model is the follow-up to the *Eightfold Way* classification scheme.

All quarks are assigned a baryon number of $\frac{1}{3}$. Up, charm and top quarks have an electric charge of $+\frac{2}{3}$, while the down, strange, and bottom quarks have an electric charge of $-\frac{1}{3}$. Antiquarks have the opposite quantum numbers. Quarks are also spin-$\frac{1}{2}$ particles, meaning they are fermions.

Mesons are made of a valence quark-antiquark pair (thus have a baryon number of 0), while baryons are made of three quarks (thus have a baryon number of 1). This article discusses the quark model for the up, down, and strange flavours of quark (which form an approximate SU(3) symmetry). There are generalizations to larger number of flavours.

The Eightfold Way classification is named after the following fact. If we take three flavours of quarks, then the quarks lie in the fundamental representation, 3 (called the triplet) of flavour SU(3). The antiquarks lie in the complex conjugate representation $\bar{3}$. The nine states (nonet) made out of a pair can be decomposed into the trivial representation, 1 (called the singlet), and the adjoint representation, 8 (called the octet). The notation for this decomposition is

$$3 \otimes \bar{3} = 8 \oplus 1$$

If the flavour symmetry were exact, then all nine mesons would have the same mass. The physical content of the theory includes consideration of the symmetry breaking induced by the quark mass differences, and considerations of mixing between various multiplets (such as the octet and the singlet). The splitting between the η and the η' is larger than the quark model can accommodate. This "η-η' puzzle" is resolved by instantons.

Richard Edward Taylor

Richard Edward Taylor, is a Canadian-American professor (Emeritus) at Stanford University. In 1990, he shared the Nobel Prize in Physics with Jerome Friedman and Henry Kendall for their pioneering investigations concerning deep inelastic scattering of electrons on protons and bound neutrons, which have been of essential importance for the development of the quark model in particle physics.

Born	2 November 1929 (age 80), Medicine Hat, Alberta
Fields	Particle physics
Institutions	SLAC, LBL, École Normale Supérieure
Alma mater	Stanford, University of Alberta
Doctoral advisor	Robert F. Mozley
Notable awards	Nobel Prize in Physics (1990)

QUARK MODEL (1990)

Developing classification schemes for hadrons became a burning question after new experimental techniques uncovered so many of them that it became clear that they could not all be elementary. These discoveries led Wolfgang Pauli to exclaim "Had I foreseen that, I would have gone into botany" (sometimes quoted as saying to Leon Lederman: Young man, if I could remember the names of these particles, I would have been a botanist), but brought a Nobel Prize for the experimental particle physicist Luis Alvarez who was at the forefront of many of these developments. Several early proposals, such as the one by Shoichi Sakata, were unable to explain all the data. A version developed by Moo-Young Han and Yoichiro Nambu was also eventually found untenable. The quark model in its modern form was developed by Murray Gell-Mann and Kazuhiko Nishijima. The model received important contributions from Yuval Ne'eman and George Zweig. The spin $\frac{3}{2}$ Ω^- baryon, a member of the ground state decuplet, was a prediction of the model. When it was discovered in an experiment at Brookhaven National Laboratory, Gell-Mann received a Nobel prize for his work on the quark model.

While the quark model is derivable from the theory of quantum chromo dynamics, the structure of hadrons is more complicated than this model reveals. The full quantum mechanics wave function of any hadron must include virtual quark pairs as well as virtual gluons. Also, there may be hadrons which lie outside the quark model. Among these are the *glueballs* (which contain only valence gluons), *hybrids* (which contain valence quarks as well as gluons) and exotic hadrons (such as tetraquarks or pentaquarks).

Mesons are hadrons with zero baryon number. If the quark-antiquark pair are in an orbital angular momentum L state, and have spin S, then

- $|L - S| \leq J \leq L + S$, where $S = 0$ or 1.
- $P = (-1)^{L+1}$, where the 1 in the exponent arises from the intrinsic parity of the quark-antiquark pair.
- $C = (-1)^{L+S}$ for mesons which have no flavour. Flavoured mesons have indefinite value of C.
- For isospin $I = 1$ and 0 states, one can define a new multiplicative quantum number called the G-parity such that $G = (-1)^{I+L+S}$.

If $P = (-1)^J$, then it follows that $S = 1$, thus $PC = 1$. States with these quantum numbers are called natural parity states while all other quantum numbers are called exotic (for example, the state $J^{PC} = 0^-$).

Pierre-Gilles de Gennes

 Pierre-Gilles de Gennes was a French physicist and the Nobel Prize laureate in Physics in 1991. His Nobel Prize was awarded for discovering that methods developed for studying order phenomena in simple systems can be generalized to more complex forms of matter, in particular to liquid crystals and polymers.

Born	24 October 1932, Paris, France
Died	18 May 2007 (age 74), Orsay, France
Nationality	French
Fields	Physics
Institutions	ESPCI, Collège de France, Paris-Sud 11 University Orsay
Alma mater	École Normale Supérieure
Notable awards	Nobel Prize in Physics (1991), Lorentz Medal(1990), Wolf Prize(1990)

GRANULAR MATERIAL (1991)

A granular material is a conglomeration of discrete solid, macroscopic particles characterized by a loss of energy whenever the particles interact (the most common example would be friction when grains collide). The constituents that compose granular material must be large enough such that they are not subject to thermal motion fluctuations. Thus, the lower size limit for grains in granular material is about 1 μm. On the upper size limit, the physics of granular materials may be applied to ice floes where the individual grains are icebergs and to asteroid belts of the solar system with individual grains being asteroids.

Some examples of granular materials are nuts, coal, sand, rice, coffee, corn flakes, fertilizer, and ball bearings. Powders are a special class of granular material due to their small particle size, which makes them more cohesive and more easily suspended in a gas. Granular materials are commercially important in applications as diverse as pharmaceutical industry, agriculture, and energy production. Research into granular materials is thus directly applicable and goes back at least to Charles-Augustin de Coulomb, whose law of friction was originally stated for granular materials.

The soldier/physicist Brigadier Ralph Alger Bagnold was an early pioneer of the physics of granular matter and whose book The Physics of Blown Sand and Desert Dunes remains an important reference to this day.

According to material scientist Patrick Richard, Granular materials are ubiquitous in nature and are the second most manipulated material in industry (the first one is water).

In some sense, Granular materials do not constitute a single phase of matter but have characteristics reminiscent of solids, liquids, or gases depending on the average energy per grain. However, in each of these states granular materials also exhibit properties which are unique.

Granular materials also exhibit a wide range of pattern forming behaviors when excited (e.g. vibrated or allowed to flow). As such granular materials under excitation can be thought of as an example for a complex system.

When the average energy of the individual grains is low and the grains are fairly stationary relative to each other, the granular material acts like a solid. In general, stress in a granular solid is not distributed uniformly but is conducted away along so called force chains which are networks of grains resting on one another. Between these chains are regions of low stress whose grains are shielded for the effects of the grains above by vaulting and arching.

Georges Charpak

Georges Charpak is a Polish-French physicist and Nobel laureate in Physics. He was made a member of the French Academy of Sciences in 1985. In 1992, he received the Nobel Prize in Physics for his invention and development of particle detectors, in particular the multiwire proportional chamber. This is the last time a single person has won the physics prize.

Born	1 August 1924 (age 85), Dabrowica, Ukraine
Nationality	Polish, French
Fields	Physics
Known for	Multiwire proportional chamber
Notable awards	Nobel Prize in Physics (1992)

WIRE CHAMBER (1992)

A multiwire chamber (or just wire chamber) is a detector for particles of ionizing radiation which is an advancement of the concept of the Geiger counter and the proportional counter.

A proportional counter uses a Geiger-Müller tube: a wire, under high voltage, runs down the length of a metal tube whose walls are held at ground potential. The tube is filled with carefully chosen gas, such that any ionizing particle that passes through the tube will ionize surrounding gaseous atoms. The resulting ions and electrons are accelerated by the potential on the wire, causing a cascade of ionization which is collected on the wire and results in an electric current. This allows the experimenter to count particles and, in the case of the proportional counter, determine their energy.

For high energy physics experiments, it is also valuable to observe the particle's path. For a long time, bubble chambers were used for this purpose, but with the improvement of electronics, it became desirable to have a detector with fast electronic read-out. A wire chamber is a chamber with many parallel wires, arranged as a grid and put on high voltage, with the metal casing being on ground potential. As in the Geiger counter, a particle leaves a trace of ions and electrons, which drift toward the case or the *nearest* wire, respectively. By marking off the wires which had a pulse of current, one can see the particle's path.

An improvement is the multiwire proportional chamber (MWPC) which combines this with the idea of the proportional counter to determine the energy. The 1968 invention of this device won Georges Charpak the 1992 Nobel Prize in Physics.

Often, the chamber is put into a homogeneous magnetic field, so that charged particles are led into spiral paths due to the Lorentz force. By checking the direction of the curves, one can see whether and how the particles are charged. The necessary magnetic fields are often quite strong: physicists at CERN like to tell visitors the story of how Charpak once was working on an MWPC, being so careless as to sit on an iron chair. He and his colleagues spent months carefully attaching thousands of thin wires. One day, he moved his chair a bit too close to the magnetic field. The magnet pulled his chair out from under him into the chamber, tearing apart all the wires and ruining the detector.

If one also precisely measures the timing of the current pulses of the wire and takes into account that the ions need some time to drift to the nearest wire, one can infer the distance at which the particle passed the wire. This greatly increases the accuracy of the path reconstruction and is known as a drift chamber.

If two drift chambers are used with the wires of one orthogonal to the wires of the other, both orthogonal to the beam direction, a more precise detection of the position is obtained. If an additional simple detector is used to detect, with poor or null positional resolution, the particle at a fixed distance before or after the wires, a tridimensional reconstruction can be made and the speed of the particle deducted from the difference in time of the passage of the particle in the different part of the detector. This setup gives up the detector called Time Projection Chamber.

Russell Alan Hulse Fowler

 Russell Alan Hulse is an American physicist and winner of the Nobel Prize in Physics, shared with his thesis advisor Joseph Hooton Taylor Jr., for the discovery of a new type of pulsar, a discovery that has opened up new possibilities for the study of gravitation. He is a specialist in the pulsar studies and gravitational waves.

Born	28 November 1950 (age 59), New York City, New York
Nationality	United States
Institutions	UT Dallas, Princeton Plasma Physics Laboratory, NRAO
Alma mater	UMass Amherst
Notable awards	Nobel Prize in Physics (1993)

GRAVITATIONAL WAVE (1993)

In physics, a gravitational wave is a fluctuation in the curvature of space time which propagates as a wave, traveling outward from the source. Predicted to exist by Albert Einstein in 1916 on the basis of his theory of general relativity, the waves transport energy known as gravitational radiation. Sources of gravitational waves include binary star systems composed of white dwarfs, neutron stars, or black holes.

Although gravitational radiation has not yet been directly detected, it has been indirectly shown to exist. This was the basis for the 1993 Nobel Prize in Physics, awarded for measurements of the Hulse-Taylor binary system. Various gravitational wave detectors exist.

PULSAR

Pulsars are highly magnetized, rotating neutron stars that emit a beam of electromagnetic radiation. The radiation can only be observed when the beam of emission is pointing towards the Earth. This is called the lighthouse effect and gives rise to the pulsed nature that gives pulsars their name. Because neutron stars are very dense objects, the rotation period and thus the interval between observed pulses is very regular. For some pulsars, the regularity of pulsation is as precise as an atomic clock. The observed periods of their pulses range from 1.4 milliseconds to 8.5 seconds. A few pulsars are known to have planets orbiting them, such as PSR B1257+12. Werner Becker of the Max Planck Institute for Extraterrestrial Physics said in 2006, The theory of how pulsars emit their radiation is still in its infancy, even after nearly forty years of work.

The events leading to the formation of a pulsar begin when the core of a massive star is compressed during a supernova, which collapses into a neutron star. The neutron star retains most of its angular momentum, and since it has only a tiny fraction of its progenitor's radius (and therefore its moment of inertia is sharply reduced), it is formed with very high rotation speed. A beam of radiation is emitted along the magnetic axis of the pulsar, which spins along with the rotation of the neutron star. The magnetic axis of the pulsar determines the direction of the electromagnetic beam, with the magnetic axis not necessarily being the same as its rotational axis. This misalignment causes the beam to be seen once for every rotation of the neutron star, which leads to the pulsed nature of its appearance. The beam originates from the rotational energy of the neutron star, which generates an electrical field from the movement of the very strong magnetic field, resulting in the acceleration of protons and electrons on the star surface and the creation of an electromagnetic beam emanating from the poles of the magnetic field. This rotation slows down over time as electromagnetic power is emitted. When a pulsar's spin period slows down sufficiently, the radio pulsar mechanism is believed to turn off (the so-called death line). This turn-off seems to take place after about 10–100 million years, which means of all the neutron stars in the 13.6 billion year age of the universe, around 99% no longer pulsate. To date, the slowest observed pulsar has a period of 8 seconds.

Joseph Hooton Taylor, Jr.

Joseph Hooton Taylor, Jr. is an American astrophysicist and winner of Nobel Prize in Physics for his discovery with Russell Alan Hulse, of a new type of pulsar, a discovery that has opened up new possibilities for the study of gravitation.

Born	29 March 1941 (age 69), Philadelphia, Pennsylvania, USA
Nationality	United States
Fields	Physics
Institutions	Princeton University, University of Massachusetts, Five College Radio Astronomy Observatory
Alma mater	Haverford College, Harvard University
Doctoral students	Victoria Kaspi
Known for	Pulsars
Notable awards	Wolf Prize in Physics (1992), Nobel Prize in Physics (1993)

PULSAR (1993)

The first pulsar was observed on 28 November 1967 by Jocelyn Bell Burnell and Antony Hewish. Initially baffled as to the seemingly unnatural regularity of its emissions, they dubbed their discovery LGM-1, for little green men. While the hypothesis that pulsars were beacons from extraterrestrial civilizations was never taken very seriously, some discussed the far-reaching implications if it turned out to be true. Their pulsar was later dubbed CP 1919, and is now known by a number of designators including PSR 1919 + 21, PSR B1919 + 21 and PSR J1921 + 2153. Although CP 1919 emits in radio wavelengths, pulsars have, subsequently, been found to emit in visible light, X-ray, and/or gamma-ray wavelengths.

The word pulsar is a contraction of pulsating star, and first appeared in print in 1968:

An entirely novel kind of star came to light on 6 August, last year and was referred to, by astronomers, as LGM (Little Green Men). Now it is thought to be a novel type between a white dwarf and a neutron. The name Pulsar is likely to be given to it. Dr. A. Hewish told me yesterday: I am sure that today every radio telescope is looking at the Pulsars.

The suggestion that pulsars were rotating neutron stars was put forth independently by Thomas Gold and Franco Pacini in 1968, and was soon proven beyond reasonable doubt by the discovery of a pulsar with a very short (33-millisecond) pulse period in the Crab nebula.

In 1974, Antony Hewish became the first astronomer to be awarded the Nobel Prize in physics. Considerable controversy is associated with the fact that Professor Hewish was awarded the prize while Bell, who made the initial discovery while she was his PhD student, was not.

The events leading to the formation of a pulsar begin when the core of a massive star is compressed during a supernova, which collapses into a neutron star. The neutron star retains most of its angular momentum, and since it has only a tiny fraction of its progenitor's radius, it is formed with very high rotation speed. A beam of radiation is emitted along the magnetic axis of the pulsar, which spins along with the rotation of the neutron star. The magnetic axis of the pulsar determines the direction of the electromagnetic beam, with the magnetic axis not necessarily being the same as its rotational axis. This misalignment causes the beam to be seen once for every rotation of the neutron star, which leads to the pulsed nature of its appearance. The beam originates from the rotational energy of the neutron star, which generates an electrical field from the movement of the very strong magnetic field, resulting in the acceleration of protons and electrons on the star surface and the creation of an electromagnetic beam emanating from the poles of the magnetic field.

Bertram Neville Brockhouse

 Bertram Neville Brockhouse, was a Canadian physicist. He was awarded the Nobel Prize in Physics (1994, shared with Clifford Shull) for pioneering contributions to the development of neutron scattering techniques for studies of condensed matter, in particular for the development of neutron spectroscopy.

Born	15 July 1918, Lethbridge, Alberta
Died	13 October 2003 (age 85), Hamilton, Ontario
Nationality	Canada
Institutions	McMaster University
Notable awards	Nobel Prize in Physics (1994)

NEUTRON SCATTERING (1994)

Neutron Scattering encompasses all scientific techniques whereby the deflection of neutron radiation is used as a scientific probe. Neutrons readily interact with atomic nuclei and magnetic fields from unpaired electrons, making a useful probe of both structure and magnetic order. Neutron Scattering falls into two basic categories: elastic and inelastic. Elastic scattering is when a neutron interacts with a nucleus or electronic magnetic field but does not leave it in an excited state, meaning the emitted neutron has the same energy as the injected neutron. Scattering processes that involve an energetic excitation or relaxation by the neutron are inelastic: the injected neutron's energy is used or increased to create an excitation or by absorbing the excess energy from a relaxation, and consequently the emitted neutron's energy is reduced or increased respectively.

For several good reasons, moderated neutrons provide an ideal tool for the study of almost all forms of condensed matter. Firstly, they are readily produced at a nuclear research reactor or a spallation source. Normally in such processes neutrons are however produced with much higher energies than are needed. Therefore, moderators are generally used which slow the neutrons down, and therefore, produce wavelengths that are comparable to the atomic spacing in solids and liquids, and kinetic energies that are comparable to those of dynamic processes in materials. Moderators can be made from aluminum and filled with liquid hydrogen (for very long wavelength neutrons) or liquid methane (for shorter wavelength neutrons). Fluxes of $10^7/s$–$10^8/s$ are not atypical in most neutron sources from any given moderator.

The neutrons cause pronounced interference and energy transfer effects in scattering experiments. Unlike an X-ray photon with a similar wavelength, which interacts with the electron cloud surrounding the nucleus, neutrons interact with the nucleus itself. Because the neutron is an electrically neutral particle, it is deeply penetrating, and is therefore, more able to probe the bulk material. Consequently, it enables the use of a wide range of sample environments that are difficult to use with synchrotron X-ray sources. It also has the advantage that the cross sections for interaction do not increase with atomic number as they do with radiation from a synchrotron X-ray source. Thus neutrons can be used to analyze materials with low atomic numbers like proteins and surfactants. This can be done at synchrotron sources but very high intensities are needed which may cause the structures to change. Moreover, the nucleus provides a very short range, isotropic potential varying randomly from isotope to isotope, making it possible to tune the nuclear scattering contrast to suit the experiment.

The neutron has an additional advantage over the X-ray photon in the study of condensed matter. It readily interacts with internal magnetic fields in the sample. In fact, the strength of the magnetic scattering signal is often very similar to that of the nuclear scattering signal in many materials, which allows the simultaneous exploration of both nuclear and magnetic structure.

Clifford Glenwood Shull

Clifford Glenwood Shull was a Nobel Prize-winning American physicist. Clifford G. Shull was awarded the 1994 Nobel Prize in Physics with Canadian Bertram Brockhouse. This is the longest ever time after the original work was completed that the Nobel Prize was awarded.

The two won the prize for the development of the neutron scattering technique. He also conducted research on condensed matter. 'Professor Shull's prize was awarded for his pioneering work in neutron scattering, a technique that reveals where atoms are within a material like ricocheting bullets reveal where obstacles are in the dark.

Born	23 September 1915, Pittsburgh, Pennsylvania
Died	31 March 2001, Medford, Massachusetts
Nationality	United States
Fields	Physics
Known for	Neutron scattering
Notable awards	Nobel Prize in Physics (1994)

SMALL-ANGLE NEUTRON SCATTERING (1994)

Small angle neutron scattering (SANS) is a laboratory technique, similar to the often complementary techniques of small angle X-ray scattering (SAXS) and light scattering. These are particularly useful because of the dramatic increase in forward scattering that occurs at phase transitions, known as critical opalescence, and because many materials, substances and biological systems possess interesting and complex features in their structure, which match the useful length scale ranges that these techniques probe. The technique provides valuable information over a wide variety of scientific and technological applications including chemical aggregation, defects in materials, surfactants, colloids, ferromagnetic correlations in magnetism, alloy segregation, polymers, proteins, biological membranes, viruses, ribosome and macromolecules. There are numerous SANS instruments available worldwide. While analysis of the data can give information on size, shape, etc., without making any model assumptions a preliminary analysis of the data can only give information on the radius of gyration for a particle using Guinier's equation.

Neutron scattering encompasses all scientific techniques whereby the deflection of neutron radiation is used as a scientific probe. Neutrons readily interact with atomic nuclei and magnetic fields from unpaired electrons, making a useful probe of both structure and magnetic order. Neutron Scattering falls into two basic categories: elastic and inelastic. Elastic scattering is when a neutron interacts with a nucleus or electronic magnetic field but does not leave it in an excited state, meaning the emitted neutron has the same energy as the injected neutron. Scattering processes that involve an energetic excitation or relaxation by the neutron are inelastic: the injected neutron's energy is used or increased to create an excitation or by absorbing the excess energy from a relaxation, and consequently the emitted neutron's energy is reduced or increased respectively.

During a SANS experiment a beam of neutrons is directed at a sample, which can be an aqueous solution, a solid, a powder, or a crystal. The neutrons are elastically scattered by changes of refractive index on a nanometer scale inside the sample which is the interaction with the nuclei of the atoms present in the sample. Because the nuclei of all atoms are compact and of comparable size neutrons are capable of interacting strongly with all atoms. This is in contrast to X-ray techniques where the X-rays interact weakly with hydrogen, the most abundant element.

Martin Lewis Perl

 Martin Lewis Perl is an American physicist, who won the Nobel Prize in Physics in 1995 for his discovery of the tau lepton. His parents were Jewish emigrants to the US from the Polish area of Russia.

Perl is a 1948 chemical engineering graduate of Brooklyn Polytechnic Institute in Brooklyn, and received his Ph.D. from Columbia University in 1955. He spent his career at the University of Michigan and then at the Stanford Linear Accelerator Center (SLAC).

Born	24 June 1927 (age 82), New York
Nationality	United States
Fields	Physics
Institutions	University of Michigan, Stanford Linear Accelerator Center (SLAC)
Alma mater	Columbia University
Known for	tau lepton
Notable awards	Nobel Prize in Physics (1995)

TAUON (1995)

The tauon was detected in a series of experiments between 1974 and 1977 by Martin Lewis Perl with his colleagues at the SLAC-LBL group. Their equipment consisted of SLACs then new e^+–e^- colliding ring, called SPEAR, and the LBL magnetic detector. They could detect and distinguish between leptons, hadrons and photons. They did not detect the tauon directly, but rather discovered anomalous events:

They have discovered 64 events of the form:

$e^+ + e^- \rightarrow e^\pm + \mu^\mp + $ at least 2 undetected particles

for which we have no conventional explanation.

The need for at least 2 undetected particles was shown by the inability to conserve energy and momentum with only one. However, no other muons, electrons, photons, or hadrons were detected. It was proposed that this event was the production and subsequent decay of a new particle pair:

$$e^+ + e^- \rightarrow \tau^+ + \tau^- \rightarrow e^\pm + \mu^\mp + 4\nu.$$

This was difficult to verify, because the energy to produce the $\tau^+\tau^-$ pair is similar to the threshold for D meson production. Work done at DESY-Hamburg, and with the Direct Electron Counter (DELCO) at SPEAR, subsequently established the mass and spin of the tau.

The symbol τ was derived from the Greek (triton, meaning third in English), since it was the third charged lepton discovered.

The tauon also called the tau, tau lepton, or tau particle, is an elementary particle similar to the electron, with negative electric charge and a spin of $\frac{1}{2}$. Together with the electron, the muon, and the three neutrinos, it is classified as a lepton. Like all elementary particles, the tauon has a corresponding antiparticle of opposite charge but equal mass and spin: the antitauon (also called the a *positive tauon*). Tauons are denoted by τ^- and antitauons by τ^+.

Tauons have a lifetime of 2.9×10^{-13} s and a mass of 1,777 Mev/c^2 (compared to 105.7 Mev/c^2 for muons and 0.511 Mev/c^2 for electrons). Since their interactions are very similar to those of the electron, a tauon can be thought of as a much heavier version of the electron. Because of their greater mass, tauons do not emit as much bremsstrahlung radiation as electrons; consequently they are potentially highly penetrating, much more so than electrons. However, because of their short lifetime, the tauons range is mainly set by their decay length, which is too small for bremsstrahlung to be noticeable: their penetrating power appears only at ultra high energy (above Pev energies).

As with the case of the other charged leptons, the tauon has an associated tauon neutrino. Tauon neutrinos are denoted by ν_τ.

Frederick Reines

Frederick Reines was an American physicist. He was awarded the 1995 Nobel Prize in Physics for his co-detection of the neutrino with Clyde Cowan in the neutrino experiment, and may be the only scientist in history so intimately associated with the discovery of an elementary particle and the subsequent thorough investigation of its fundamental properties.

Born	16 March 1918, Paterson, New Jersey
Died	26 August 1998 (age 80)
Citizenship	United States
Fields	Physics
Known for	Neutrinos
Notable awards	Nobel Prize in Physics (1995), Franklin Medal (1992)

EXPERIMENTAL DEMONSTRATION OF NEUTRINO FLAVORS (1995)

The first detection of tau neutrino interactions was announced in summer of 2000 by the DONUT collaboration at Fermi lab, making it the latest particle of the Standard Model to have been directly observed; its existence had already been inferred by both theoretical consistency and experimental data from the Large Electron-Positron Collider.

Starting in the late 1960s, several experiments found that the number of electron neutrinos arriving from the Sun was between one third and one half the number predicted by the Standard Solar Model (SSM). This discrepancy, which became known as the solar neutrino problem, remained unresolved for some thirty years. The Standard Model of particle physics (SM) assumes that neutrinos are mass less and cannot change flavor. However, if neutrinos had mass, they could change flavor (or *oscillate* between flavours).

A practical method for investigating neutrino oscillations was first suggested by Bruno Pontecorvo in 1957 using an analogy with Kaon oscillations; over the subsequent 10 years he developed the mathematical formalism and the modern formulation of vacuum oscillations. In 1985 Stanislav Mikheyev and Alexei Smirnov (expanding on 1978 work by Lincoln Wolfenstein) noted that flavor oscillations can be modified when neutrinos propagate through matter. This so-called Mikheyev-Smirnov-Wolfenstein effect (MSW effect) is important to understand because many neutrinos emitted by fusion in the Sun pass through the dense matter in the solar core (where essentially all solar fusion takes place) on their way to detectors on Earth.

Direct detection of flavor oscillation in solar neutrinos

Starting in 1998, experiments began to show that solar and atmospheric neutrinos change flavours. This resolved the solar neutrino problem: the electron neutrinos produced in the Sun had partly changed into other flavours which the experiments could not detect.

Although individual experiments, such as the set of solar neutrino experiments, are consistent with non-oscillatory mechanisms of neutrino flavour conversion, taken altogether, neutrino experiments imply the existence of neutrino oscillations. Especially relevant in this context are the reactor experiment KamLAND and the accelerator experiments such as MINOS. The KamLAND experiment has indeed identified oscillations as the neutrino flavour conversion mechanism involved in the solar electron neutrinos. Similarly, MINOS confirms the oscillation of atmospheric neutrinos and gives a better determination of the mass squared splitting.

David Morris Lee

 David Morris Lee is an American physicist who shared the 1996 Nobel Prize in Physics with Robert C. Richardson and Douglas Osheroff for their discovery of super fluidity in helium-3. The work that led to Lee's Nobel Prize was performed in the early 1970s.

Lee, together with Robert C. Richardson and graduate student, Doug Osheroff used a Pomeranchuk cell to investigate the behaviour of ^{3}He at temperatures within a few thousands of a degree of absolute zero. They discovered unexpected effects in their measurements, which they eventually explained as phase transitions to a superfluid phase of ^{3}He. Lee, Richardson and Osheroff were jointly awarded the Nobel Prize in Physics in 1996 for this discovery.

Born	20 January 1931 (age 79), Rye, New York
Fields	Physics
Institutions	Cornell University, Texas A&M University (2009-present)
Alma mater	Yale University, University of Connecticut, Harvard University
Doctoral advisor	Henry A. Fairbank
Notable awards	Nobel Prize in Physics (1996), Oliver Buckley Prize (1981), Sir Francis Simon Memorial Prize (1976)

HELIUM-3 (1996)

Helium-3 (He-3) is a light, non-radioactive isotope of helium with two protons and one neutron. It is rare on Earth, and is sought for use in nuclear fusion research. The abundance of helium-3 is thought to be greater on the Moon (embedded in the upper layer of regolith by the solar wind over billions of years) and the solar system's gas giants (left over from the original solar nebula), though still low in quantity (28 ppm of lunar regolith is helium-4 and 0.01 ppm is helium-3).

The helion, the nucleus of a helium-3 atom, consists of two protons but only one neutron, in contrast to two neutrons in ordinary helium. Its existence was first proposed in 1934 by the Australian nuclear physicist Mark Oliphant while based at Cambridge University's Cavendish Laboratory, in an experiment in which fast deuterons were reacted with other deuteron targets (the first demonstration of nuclear fusion). Helium-3, as an isotope, was postulated to be radioactive, until helions from it were accidentally identified as a trace contaminant in a sample of natural helium (which is mostly helium-4) from a gas well, by Luis W. Alvarez and Robert Cornog in a cyclotron experiment at the Lawrence Berkeley National Laboratory, in 1939. The presence of helium-3 in underground gas deposits implied that it either did not decay or had an extremely long half-life compatible with a primordial isotope.

Helium-3 is proposed as a second-generation fusion fuel for fusion power uses. Tritium, with a 12-year half-life, decays into helium-3, which can be recovered. Irradiation of lithium in a nuclear reactor either a fusion or fission reactor can also produce tritium, and thus (after decay) helium-3.

Helium-3 is a most important isotope in instrumentation for neutron detection. It has a high absorption cross-section for thermal neutron beams and is used as a converter gas in neutron detectors. The neutron is converted through the nuclear reaction

$$n + {}^3He \rightarrow {}^3H + {}^1H + 0.764 \; Mev$$

into charged particles tritium (T, ^3H) and protium (p, ^1H) which then are detected by creating a charge cloud in the stopping gas of a proportional counter or a Geiger-Müller tube.

Douglas Dean Osheroff

Douglas Dean Osheroff is an American physicist who shared the 1996 Nobel Prize in Physics with David Lee and Robert C. Richardson for their discovery of super fluidity in helium-3.

Osheroff used a Pomeranchuk cell to investigate the behaviour of ^{3}He at temperatures within a few thousands of a degree of absolute zero. They discovered unexpected effects in their measurements, which they eventually explained as phase transitions to a superfluid phase of ^{3}He.

Born	1 August 1945 (age 64), Aberdeen, Washington
Nationality	United States
Fields	Physics
Institutions	Bell Labs, Stanford University
Alma mater	California Institute of Technology (B.S.), Cornell University (PhD)
Known for	Discovering superfluidity in helium-3
Notable awards	Nobel Prize in Physics (1996)

HELIUM-3 (1996)

Physical properties: Due to the lower atomic mass of helium-3 (3.0160293 amu), it has significantly different properties from helium-4 (4.0026 amu). Because of the weak induced dipole-dipole interaction between helium atoms, their physical properties are mainly determined by zero point energy, and the lower mass of helium-3 causes it to have higher zero point energy, which means helium-3 can overcome dipole-dipole interaction with less thermal energy than helium-4. Helium-3 boils at 3.19 Kelvin compared to helium-4s 4.23 K, and its critical point is also lower at 3.35 K, compared to helium-4s 5.19 K. It has less than half the density when liquid at its boiling point: 0.059 g/ml compared to helium-4s 0.12473 g/ml at one atmosphere. Its latent heat of vaporization is also considerably lower at 0.026 kJ/mol compared to helium-4s 0.0829 kJ/mol.

Fusion reactions: The fusion reaction rate increases rapidly with temperature until it maximizes and then gradually drops off. The DT rate peaks at a lower temperature (about 70 keV, or 800 million Kelvin's) and at a higher value than other reactions commonly considered for fusion energy.

Some fusion processes produce highly energetic neutrons which render reactor components radioactive with activation products through the continuous bombardment of the reactor's components with emitted neutrons. Because of this bombardment and irradiation, power generation must occur indirectly through thermal means, as in a fission reactor. However, the appeal of helium-3 fusion stems from the aneutronic nature of its reaction products. Helium-3 itself is non-radioactive. The lone high-energy by-product, the proton, can be contained using electric and magnetic fields. The momentum energy of this proton will interact with the containing electromagnetic field, resulting in direct net electricity generation.

Due to the higher Coulomb barrier, the temperatures required for $^2_1H + ^3_2He$ fusion are much higher than those of conventional $^2_1H + ^3_1H$ (deuterium + tritium) fusion. Moreover, since both reactants need to be mixed together to fuse, reactions between nuclei of the same reactant will occur, and the D-D reaction ($^2_1H + ^2_1H$) does produce a neutron. Reaction rates vary with temperature, but the D-^{3}He reaction rate is never greater than 3.56 times the D-D reaction rate. Therefore fusion using D-^{3}He fuel may produce a somewhat lower neutron flux than D-T fusion, but is by no means clean, negating some of its main attraction.

A second possibility, fusing ^3_2He with itself ($^3_2He + ^3_2He$), requires even higher temperatures (since now both reactants have a +2 charge), and thus is even more difficult than the D-^{3}He reaction. However, it does offer a possible reaction that produces no neutrons.

Robert Coleman Richardson

Robert Coleman Richardson is an American experimental physicist whose area of research includes sub-millikelvin temperature studies of helium-3.

Richardson, along with David Lee, as senior researchers, and then graduate student Douglas Osheroff, shared the 1996 Nobel Prize in Physics for their 1972 discovery of the property of super fluidity in helium-3 atoms in the Cornell University Laboratory of Atomic and Solid State Physics.

Born	26 June 1937 (age 72), Washington D.C.
Nationality	United States
Residence	United States
Fields	Physics
Institutions	Cornell University
Alma mater	Virginia Tech, Duke University
Doctoral advisor	Horst Meyer
Known for	Discovering super fluidity in helium-3
Notable awards	Nobel Prize in Physics (1996)

SUPER FLUID (1996) / HELIUM-3

Although the phenomenologies of the super fluid states of helium-4 and helium-3 are very similar, the microscopic details of the transitions are very different. Helium-4 atoms are bosons, and their super fluidity can be understood in terms of the Bose statistics that they obey. Specifically, the super fluidity of helium-4 can be regarded as a consequence of Bose-Einstein condensation in an interacting system. On the other hand, helium-3 atoms are fermions, and the super fluid transition in this system is described by a generalization of the BCS theory of superconductivity. In it, Cooper pairing takes place between atoms rather than electrons, and the attractive interaction between them is mediated by spin fluctuations rather than phonons. A unified description of superconductivity and super fluidity is possible in terms of gauge symmetry breaking.

Super fluids, such as super cooled helium-4, exhibit many unusual properties. Super fluid acts as if it were a mixture of a normal component, with all the properties associated with normal fluid, and a super fluid component. The super fluid component has zero viscosity, zero entropy, and infinite thermal conductivity. It is thus impossible to set up a temperature gradient in a super fluid, much as it is impossible to set up a voltage difference in a superconductor. Application of heat to a spot in super fluid helium results in a wave of heat conduction at the relatively high velocity of 20 m/s, called second sound.

One of the most spectacular results of these properties is known as the thermo mechanical or fountain effect. If a capillary tube is placed into a bath of super fluid helium and then heated, even by shining a light on it, the super fluid helium will flow up through the tube and out the top as a result of the Clausius-Clapeyron relation. A second unusual effect is that super fluid helium can form a layer, 30 nm thick, up the sides of any container in which it is placed.

A more fundamental property than the disappearance of viscosity becomes visible if super fluid is placed in a rotating container. Instead of rotating uniformly with the container, the rotating state consists of quantized vortices. That is, when the container is rotated at speed below the first critical velocity the liquid remains perfectly stationary. Once the first critical velocity is reached, the super fluid will very quickly begin spinning at the critical speed. The speed is quantized, that is, a super fluid can only spin at certain allowed or critical speed values. In simplified terms, if the container is rotated to a certain allowed speed, the super fluid will rotate very quickly along with the container, otherwise, if the speed is too slow, then the super fluid will not move at all. Rotation in a normal fluid like water is not quantized.

Steven Chu

 Steven Chu is an American physicist and currently the 12th United States Secretary of Energy. Working at Bell Labs and Stanford University, Chu is known for his research in cooling and trapping of atoms with laser light, which won him the Nobel Prize in Physics in 1997.

At the time of his appointment as Energy Secretary, he was a professor of physics and molecular and cellular biology at the University of California, Berkeley and the director of the Lawrence Berkeley National Laboratory, where his research was concerned primarily with the study of biological systems at the single molecule level. He is a vocal advocate for more research into alternative energy and nuclear power, arguing that a shift away from fossil fuels is essential in combating Climate Change. For example, he has conceived of a global glucose economy, a form of a low-carbon economy, in which glucose from tropical plants is shipped around like oil is today.

Born	28 February 1948 (age 62), St. Louis, Missouri
Fields	Experimental physics
Alma mater	University of Rochester (B.A./B.S.), University of California, Berkeley (PhD)
Notable awards	Nobel Prize in Physics (1997)

LASER COOLING (1997)

Laser cooling refers to the number of techniques in which atomic and molecular samples are cooled through the interaction with one or more laser light fields. The first example of laser cooling, and also still the most common method of laser cooling (so much so that it is still often referred to as laser cooling) is Doppler cooling.

Doppler cooling, which is usually accompanied by a magnetic trapping force to give a magneto-optical trap, is by far the most common method of laser cooling. It is used to cool low density gasses down to the Doppler cooling limit, which for Rubidium 85 is around 150 microkelvin. As Doppler cooling requires a very particular energy level structure, known as a closed optical loop, the method is limited to a small handful of elements.

In Doppler cooling, the frequency of light is tuned slightly below an electronic transition in the atom. Because the light is detuned to the red (i.e. at lower frequency) of the transition, the atoms will absorb more photons if they move towards the light source, due to the Doppler Effect. Thus if one applies light from two opposite directions, the atoms will always scatter more photons from the laser beam pointing opposite to their direction of motion. In each scattering event the atom loses a momentum equal to the momentum of the photon. If the atom, which is now in the excited state, emits a photon spontaneously, it will be kicked by the same amount of momentum but in a random direction. The result of the absorption and emission process is to reduce the speed of the atom, provided its initial speed is larger than the recoil velocity from scattering a single photon. If the absorption and emission are repeated many times, the mean velocity, and therefore the kinetic energy of the atom will be reduced. Since the temperature of an ensemble of atoms is a measure of the random internal kinetic energy, this is equivalent to cooling the atoms.

Several, somewhat similar processes are also referred to as *laser cooling*, in which photons are used to pump heat away from a material and thus cool it. The phenomenon has been demonstrated via anti-Stokes fluorescence, and both electroluminescent up conversion and photo luminescent up conversion have been studied as means to achieve the same effects. In many of these, the coherence of the laser light is not essential to the process, but lasers are typically used to achieve a high irradiance.

Claude Cohen-Tannoudji

 Claude Cohen-Tannoudji is a French physicist and Nobel Laureate. He shared the 1997 Nobel Prize in Physics for research in methods of cooling and trapping atoms using laser light. He is still an active researcher, working at the École Normale Supérieure in Paris.

Born	1 April 1933 (age 77), Constantine, Algeria
Nationality	France
Fields	Physics
Institutions	École Normale Supérieure
Notable awards	Nobel Prize in Physics (1997)

SISYPHUS COOLING (1997)

Sisyphus cooling is a mechanism through which atoms can be cooled using laser beams below the temperatures expected to be achieved by Doppler cooling. It comes about as a result of a polarization gradient created by two counter-propagating laser beams with orthogonal polarization. Atoms moving through the potential landscape created by the standing wave (created by the interference of the two counter-propagating beams) lose kinetic energy as they move to a potential maximum, at which point optical pumping moves them to a lower-energy state, thus losing the potential energy they had.

Repeated cycles of converting kinetic energy to potential energy, then losing potential energy via optical pumping allow the atoms to reach temperatures orders of magnitude below those available through simple Doppler cooling.

Sympathetic cooling:

Sympathetic cooling is a process in which particles of one type cool particles of another type.

Typically, atomic ions that can be directly laser cooled are used to cool nearby ions or atoms, by way of their mutual Coulomb interaction. This technique allows cooling of ions and atoms that can't be cooled directly by laser cooling. This includes most molecular ion species, especially large organic molecules. However, sympathetic cooling is most efficient when the mass/charge ratios of the sympathetic- and laser-cooled ions are similar.

The cooling of neutral atoms in this manner was first demonstrated by Christopher Myatt et al. in 1997. Here, a technique with electric and magnetic fields were used, where atoms with spin in one direction were more weakly confined than those with spin in the opposite direction. The weakly confined atoms with a high kinetic energy were allowed to more easily escape, lowering the total kinetic energy, resulting in a cooling of the strongly confined atoms. Myatt et al. also showed the utility of their version of sympathetic cooling for the creation of Bose-Einstein condensates.

The frequency of light is tuned slightly below an electronic transition in the atom. Because the light is detuned to the red (i.e. at lower frequency) of the transition, the atoms will absorb more photons if they move towards the light source, due to the Doppler Effect. Thus if one applies light from two opposite directions, the atoms will always scatter more photons from the laser beam pointing opposite to their direction of motion. In each scattering event the atom loses a momentum equal to the momentum of the photon. If the atom, which is now in the excited state, emits a photon spontaneously, it will be kicked by the same amount of momentum but in a random direction. The result of the absorption and emission process is to reduce the speed of the atom, provided its initial speed is larger than the recoil velocity from scattering a single photon. If the absorption and emission are repeated many times, the mean velocity, and therefore the kinetic energy of the atom will be reduced. Since the temperature of an ensemble of atoms is a measure of the random internal kinetic energy, this is equivalent to cooling the atoms.

William Daniel Phillips

William Daniel Phillips is an American physicist. He is of Italian and Welsh extraction and a Methodist. Phillips' doctoral thesis concerned magnetic moment of the proton in H_2O.

This led to connections that would be important later in his research. He later did some work with Bose-Einstein condensate. In 1997 he won the Nobel Prize in Physics together with Claude Cohen-Tannoudji and Steven Chu for his contributions to laser cooling, a technique to slow the movement of gaseous atoms in order to better study them, at the National Institute of Standards and Technology.

Born	5 November 1948 (age 61), Wilkes-Barre, Pennsylvania
Nationality	United States
Fields	Physics
Institutions	NIST, University of Maryland, College Park
Alma mater	MIT, Juniata College
Known for	Laser cooling
Notable awards	Nobel Prize in Physics (1997)

RESOLVED SIDEBAND COOLING (1997)

Resolved sideband cooling is a laser cooling technique that can be used to cool strongly trapped atoms to the quantum ground state of their motion. The atoms are usually precooled using the Doppler laser cooling. Subsequently the resolved sideband cooling is used to cool the atoms beyond the Doppler cooling limit.

A cold trapped atom can be treated to a good approximation as a quantum mechanical harmonic oscillator. If the spontaneous decay rate is much smaller than the vibrational frequency of the atom in the trap, the energy levels of the system can be resolved as consisting of internal levels each corresponding to a ladder of vibrational states.

The vast majority of photons that come anywhere near a particular atom are almost completely unaffected by that atom. The atom is almost completely transparent to most frequencies (colours) of photons.

A few photons happen to resonate with the atom, in a few very narrow bands of frequencies (a single colour rather than a mixture like white light). When one of those photons comes close to the atom, the atom typically absorbs that photon (absorption spectrum) for a brief period of time, and then emits an identical photon (emission spectrum) in some random, unpredictable direction. (Other sorts of interactions between atoms and photons exist, but are not relevant to this article.)

The popular idea that lasers increase the thermal energy of matter is not the case when examining individual atoms. If a given atom is practically motionless (a cold atom), and the frequency of a laser focused upon it can be controlled, most frequencies do not affect the atom, it is invisible at those frequencies. There are only a few points of electromagnetic frequency that have any effect on that atom. At those frequencies, the atom can absorb a photon from the laser, while transitioning to an excited electronic state, and pick up the momentum of that photon. Since the atom now has the photon's momentum, the atom must begin to drift in the direction the photon was traveling. A short time later, the atom will spontaneously emit a photon in a random direction, as it relaxes to a lower electronic state. If that photon is emitted in the direction of the original photon, the atom will give up its momentum to the photon and will become motionless again. If the photon is emitted in the opposite direction, the atom will have to provide momentum in that opposite direction, which means the atom will pick up even more momentum in the direction of the original photon (to conserve momentum), with double its original velocity. But usually the photon speeds away in some other direction, giving the atom at least some sideways thrust.

Robert Betts Laughlin

 Robert Betts Laughlin is a professor of Physics and Applied Physics at Stanford University. Along with Horst L. Störmer of Columbia University and Daniel C. Tsui of Princeton University, he was awarded the 1998 Nobel Prize in physics for his explanation of the fractional quantum Hall effect.

Laughlin was born in Visalia, California. He earned a B.A. in Mathematics from UC Berkeley in 1972, and his PhD in physics in 1979 at MIT, Cambridge, Massachusetts, USA.

Born	1 November 1950 (age 59), Visalia, California, USA
Nationality	United States
Fields	Theoretical Physics
Institutions	Stanford
Alma mater	MIT, UC Berkeley
Known for	Quantum hall effect
Notable awards	Nobel Prize in Physics (1998), The Franklin Medal (1998)

FRACTIONAL QUANTUM HALL EFFECT (1998)

The fractional quantum Hall effect (FQHE) is a physical phenomenon in which a certain system behaves as if it were composed of particles with charge smaller than the elementary charge. Its discovery and explanation were recognized by the 1998 Nobel Prize in Physics.

The fractional quantum Hall effect (FQHE) is defined as a manifestation of simple collective behavior in a two-dimensional system of strongly interacting electrons. At particular magnetic fields, the electron gas condenses into a remarkable state with liquid-like properties. This state is very delicate, requiring high quality material with a low carrier concentration, and extremely low temperatures. As in the integer quantum Hall Effect, a series of plateaus forms in the Hall resistance. Each particular value of the magnetic field corresponds to a filling factor $v = p/q$

where p and q are integers with no common factors. Here q turns out to be an odd number with the exception of two filling factors 5/2 and 7/2. The principal series of such fractions are1/3, 2/5, 3/7 etc., and 2/3, 3/5, 4/7, etc.

There were two major steps in the theory of the FQHE.

- Fractionally-charged quasiparticles: this theory, proposed by Laughlin, is based on accurate trial wave functions for the ground state at fraction $1/q$ as well as its quasiparticle and quasihole excitations. The excitations have fractional charge of magnitude $e^* = e/q$.

- **Composite fermions:** this theory was proposed by Jain, and further extended by Halperin, Lee and Read. As a result of the repulsive interactions, two (or, in general, an even number) flux quanta h/e are captured by each electron, forming integer-charged quasiparticles called composite fermions. The fractional states of electrons are understood as the integer QHE of composite fermions. This makes electrons at a filling factor 1/3, for example, behave in the same way as at filling factor 1. A remarkable result is that the filling factor 1/2 corresponds to zero magnetic fields for composite fermions, resulting in their Fermi Sea. Experiments support composite fermions theory.

The FQHE was experimentally discovered in 1982 by Daniel Tsui and Horst Störmer, in experiments performed on gallium arsenide heterostructures developed by Arthur Gossard. Tsui, Störmer, and Laughlin were awarded the 1998 Nobel Prize for their work.

Laughlin's original plasma model was extended to other fractionally charged systems by MacDonald and others. The approach based on composite fermions by Jain has now emerged as a basic paradigm encompassing most of the earlier approaches.

Horst Ludwig Störmer

Horst Ludwig Störmer is a German physicist who shared the 1998 Nobel Prize in Physics with Daniel Tsui and Robert Laughlin. The three shared the prize for their discovery of a new form of quantum fluid with fractionally charged excitations (the fractional quantum Hall effect).

He and Tsui were working at Bell Labs at the time of the experiment cited by the Nobel committee; though the experiment itself was carried out in a laboratory at the Massachusetts Institute of Technology (Laughlin did not participate in the experiment but was later able to explain its results). Störmer studied physics at the J.W. Goethe-Universität at Frankfurt am Main and completed a PhD at the University of Stuttgart in 1977. After working at Bell Labs for 20 years, he became the I.I. Rabi professor of physics and applied physics at Columbia University in New York. Perhaps as important as the work for which he won the Nobel Prize is his invention of modulation doping, a method for making extremely high mobility two dimensional electron systems in semiconductors. This enabled the later observation of the fractional quantum Hall effect.

Born	6 April 1949 (age 61), Frankfurt, Germany
Nationality	Germany
Fields	Physics
Known for	Fractional quantum Hall effect
Notable awards	Nobel Prize in Physics (1998), The Benjamin Franklin Medal (1998)

FRACTIONALLY-CHARGED QUASIPARTICLES (1998)

Apart from the FQHE itself, further evidence has continued to emerge that specifically supports the understanding that there are fractionally charged quasiparticles in an electron gas under FQHE conditions.

In 1995, the fractional charge of Laughlin quasiparticles was measured directly in a quantum antidot electrometer at Stony Brook University, New York. In 1997, two groups of physicists at the Weizmann Institute of Science in Rehovot, Israel, and at the Commissariat à l'énergie atomique laboratory near Paris, detected such quasiparticles carrying an electric current, through measuring quantum shot noise.

Symmetry breaking in physics describes a phenomenon where (infinitesimally) small fluctuations acting on a system crossing a critical point decide a system's fate, by determining which branch of a bifurcation is taken. For an outside observer unaware of the fluctuations (the noise), the choice will appear arbitrary. This process is called symmetry breaking, because such transitions usually bring the system from a disorderly state into one of two states. Since disorder is more symmetric in the sense that small variations to it do not change its overall appearance, the symmetry gets broken. Symmetry breaking is supposed to play a major role in pattern formation.

In particular, we can distinguish between:

- An explicit symmetry breaking happens when the laws describing a system are themselves not invariant under the symmetry in question.

- Spontaneous symmetry breaking describes the case where the laws are invariant but it appears the system isn't because the background of the system, its vacuum, is noninvariant. Such a symmetry breaking is parametrized by an order parameter. A special case of this type of symmetry breaking is dynamical symmetry breaking.

Daniel Chee Tsui

 Daniel Chee Tsui is a Chinese born American physicist whose areas of research included electrical properties of thin films and micro-structures of semiconductors and solid-state physics.

He is currently Arthur LeGrand Doty Professor of Electrical Engineering at Princeton University and adjunct senior research scientist in the Department of Physics at Columbia University, where he was a visiting professor from 2006 to 2008. In 1998, along with Horst L. Störmer of Columbia and Robert Laughlin of Stanford, Tsui was awarded the Nobel Prize in Physics for his contributions to the discovery of the fractional quantum Hall effect.

Born	28 February 1939 (age 71), Henan Province, China,
Residence	New Jersey, USA
Nationality	United States
Fields	Experimental physics, Electrical engineering
Institutions	Princeton University, Bell Laboratories
Alma mater	University of Chicago (PhD), Augustana College (BSc)
Known for	Quantum hall effect
Notable awards	Nobel Prize in Physics (1998)

IMPACT OF FRACTIONAL QUANTUM HALL EFFECT (1998)

The FQH effect shows the limits of Landau's symmetry breaking theory. Previously it was long believed that the symmetry breaking theory could explain all the important concepts and essential properties of all forms of matter. According to this view the only thing to be done is to apply the symmetry breaking theory to all different kinds of phases and phase transitions. From this perspective, we can understand the importance of the FQHE discovered by Tsui, Stormer, and Gossard.

Different FQH states all have the same symmetry and cannot be described by symmetry breaking theory. Thus FQH states represent new states of matter that contain a completely new kind of order—topological order. The existence of FQH liquids indicates that there is a whole new world beyond the paradigm of symmetry breaking, waiting to be explored. The FQH effect opened up a new chapter in condensed matter physics. The new types of orders represented by FQH states greatly enrich our understanding of quantum phases and quantum phase transitions. The associated fractional charge, fractional statistics, non-Abelian statistics, chiral edge states, etc demonstrate the power and the fascination of emergence in many-body systems.

In physics, a quantum phase transition (QPT) is a phase transition between different quantum phases (phases of matter at zero temperature). Contrary to classical phase transitions, quantum phase transitions can only be accessed by varying a physical parameter–such as magnetic field or pressure, at absolute zero temperature. The transition describes an abrupt change in the ground state of a many-body system due to its quantum fluctuations. Such quantum phase transitions can be first-order phase transition or continuous.

The explanation of this effect is open to new relations and interpretations of experiments in light of new theories. It is possible to consider that the FQHE really gives further evidence of the existence of magnetic flux quanta as detected in superconductivity. There are electromagnetic quanta of magnetic flux and electric elementary charge. It was shown that these fundamental quanta are determined by a physical geometric unified theory. Each particle carries an integer charge and an integer flux. Classically, if a particle crosses a line in a plane normal to its flux, there is a fractional relation between the charge and flux crossing the line and a corresponding fractional relation between the current and induced voltage if there are no resistive losses.

Gerardus (Gerard) 't Hooft

 Gerardus (Gerard) 't Hooft is a theoretical physicist at Utrecht University, the Netherlands. He shared the 1999 Nobel Prize in Physics with Martinus J.G. Veltman for elucidating the quantum structure of electroweak interactions.

Asteroid 9491 Thooft is named in his honor; he has written a constitution for its future inhabitants. He was awarded the Lorentz Medal in 1986 and the Spinozapremie in 1995. Nobel Prize in Physics laureate Frits Zernike was his great-uncle. He is married to Albertha Schik (Betteke) and has two daughters, Saskia and Ellen. Saskia has translated one of her father's popular Dutch fiction books Planetenbiljart into English. The book's title is Playing with Planets and was launched in Singapore in November 2008.

Born	5 July 1946 (age 63), Den Helder, Netherlands
Nationality	Dutch
Fields	Theoretical physics
Institutions	Utrecht University,
Alma mater	Utrecht University
Doctoral advisor	Martinus J. G. Veltman
Doctoral students	Robbert Dijkgraaf and Herman Verlinde
Known for	Quantum field theory, Quantum gravity
Notable awards	Wolf Prize (1981), Lorentz Medal (1986), Spinozapremie (1995) The Franklin Medal (1995), Nobel Prize in Physics (1999)

QUANTUM FIELD THEORY (1999)

Quantum field theory (QFT) provides a theoretical framework for constructing quantum mechanical models of systems classically parameterized by an infinite number of dynamical degrees of freedom, that is, fields and many-body systems. It is the natural and quantitative language of particle physics and condensed matter physics. Most theories in modern particle physics, including the Standard Model of elementary particles and their interactions, are formulated as relativistic quantum field theories. Quantum field theories are used in many contexts, elementary particle physics being the most vital example, where the particle count/number going into a reaction fluctuates and changes, differing from the count/number going out, for example, and for the description of critical phenomena and quantum phase transitions, such as in the BCS theory of superconductivity, quantum phase transition, critical phenomena. Quantum field theory is thought by many to be the unique and correct outcome of combining the rules of quantum mechanics with special relativity. In perturbative quantum field theory, the forces between particles are mediated by other particles. The electromagnetic force between two electrons is caused by an exchange of photons. Intermediate vector bosons mediate the weak force and gluons mediate the strong force. There is currently no complete quantum theory of the remaining fundamental force, gravity, but many of the proposed theories postulate the existence of a graviton particle that mediates it. These force-carrying particles are virtual particles and, by definition, cannot be detected while carrying the force, because such detection will imply that the force is not being carried. In addition, the notion of force mediating particle comes from perturbation theory, and thus does not make sense in a context of bound states. In QFT photons are not thought of as little billiard balls, they are considered to be field quanta—necessarily chunked ripples in a field, or excitations that look like particles.

Martinus Justinus Godefriedus Veltman

 Martinus Justinus Godefriedus Veltman, is a Dutch theoretical physicist. He shared the 1999 Nobel Prize in physics with his former student Gerardus't Hooft for their work on particle theory.

In 1981, Veltman left Utrecht University for the University of Michigan-Ann Arbor, frustrated by the recognition his student 't Hooft got for his PhD thesis. Veltman felt that he had done most of the preliminary work and written the program which made the dissertation possible. However, most of the credit went to 't Hooft. But eventually, in 1999, he was awarded the Nobel Prize for Physics in 1999 together with 't Hooft, for elucidating the quantum structure of electroweak interactions in physics. Veltman and 't Hooft joined in the celebrations at Utrecht University when the prize was awarded.

Born	27 June 1931 (age 78), Waalwijk, Netherlands
Nationality	Netherlands
Fields	Physics
Institutions	University of Michigan-Ann Arbor, Utrecht University
Alma mater	Utrecht University
Doctoral students	Gerardus 't Hooft, Peter Van Nieuwenhuizen Bernard de Wit
Notable awards	Nobel Prize in Physics (1999)

QUANTUM GRAVITY (1999)

Quantum gravity (QG) is the field of theoretical physics attempting to unify quantum mechanics with general relativity in a self-consistent manner, or more precisely, to formulate a self-consistent theory which reduces to ordinary quantum mechanics in the limit of weak gravity (potentials much less than c^2) and which reduces to Einsteinian general relativity in the limit of large actions (action much larger than reduced Planck's constant). The theory must be able to predict the outcome of situations where both quantum effects and strong-field gravity are important (at the Planck scale, unless extra dimensional theories are correct). Motivation for quantizing gravity comes from the remarkable success of the quantum theories of the other three fundamental interactions. Although some quantum gravity theories such as string theory and other so-called theories of everything attempt to unify gravity with the other fundamental forces, others such as loop quantum gravity make no such attempt; they simply quantize the gravitational field while keeping it separate from the other forces.

Observed physical phenomena in the early 21st century can be described well by quantum mechanics or general relativity, without needing both. This can be thought of as due to an extreme separation of scales at which they are important. Quantum effects are usually important only for the very small, that is, for objects no larger than typical molecules. General relativistic effects, on the other hand, show up only for the very large bodies such as collapsed stars. Planets' gravitational fields, as of 2009, are well-described by linearized gravity; so strong-field effects any effects of gravity beyond lowest nonvanishing order in ϕ/c^2 have not been observed even in the gravitational fields of planets and main sequence stars. Classical physics seems to be adequate over an enormous range of masses of objects from about 10^{-23} to 10^{30} kg. Thus there is a lack of experimental evidence relating to quantum gravity, but the gap spans 53 orders of magnitude of mass.

Zhores Ivanovich Alferov

Zhores Ivanovich Alferov, is a Russian physicist and academic who contributed significantly to the creation of modern heterostructure physics and electronics. He is an inventor of the hetero-transistor and the winner of 2000 Nobel Prize in Physics.

He is also a Russian politician and has been a member of the Russian State Parliament, the Duma, since 1995. Lately, he has become one of the most influential members of the Communist Party of the Russian Federation.

Born	15 March 1930 (age 80), Vitebsk, BSSR, Soviet Union
Nationality	Russian
Fields	Applied physics
Institutions	Ioffe Physicotechnical Institute
Alma mater	V. I. Ulyanov Electrotechnical Institute
Known for	Heterotransistors
Notable awards	Kyoto Prize in Advanced Technology (2001), Nobel Prize in Physics (2000), Demidov Prize (1999), Ioffe Prize (Russian Academy of Sciences, 1996) USSR State Prize (1984), Lenin Prize (1972)

DRIFT-FIELD TRANSISTOR (2000)

The drift-field transistor, also called the drift transistor or graded base transistor, is a type of high-speed bipolar junction transistor having a doping-engineered electric field in the base to reduce the charge carrier base transit time.

Invented by Herbert Kroemer at the Central Bureau of Telecommunications Technology of the German Postal Service, in 1953, it continues to influence the design of modern high-speed bipolar junction transistors.

HETEROJUNCTION

A heterojunction is the interface that occurs between two layers or regions of dissimilar crystalline semiconductors. These semiconducting materials have unequal band gaps as opposed to a homojunction. It is often advantageous to engineer the electronic energy bands in many solid state device applications including semiconductor lasers, solar cells and transistors (heterotransistors) to name a few. The combination of multiple heterojunctions together in a device is called a heterostructure although the two terms are commonly used interchangeably. The requirement that each material be a semiconductor with unequal band gaps is somewhat loose especially on small length scales where electronic properties depend on spatial properties. A more modern definition may be to say that a heterojunction is the interface between any two solid state materials including crystalline and amorphous structures of metallic, insulating, fast ion conductor and semiconducting material.

In 2000, the physics Nobel Prize was awarded with one half jointly to Herbert Kroemer (University of California at Santa Barbara, California, USA) and Zhores I. Alferov (A.F. Ioffe Physicotechnical Institute, St. Petersburg, Russia) for developing semiconductor heterostructures used in high-speed- and optoelectronics.

The heterojunction bipolar transistor (HBT) is an improvement of the bipolar junction transistor (BJT) that can handle signals of very high frequencies up to several hundred GHz. It is common in modern ultrafast circuits, mostly radio-frequency (RF) systems, as well as applications requiring a high power efficiency, such as power amplifiers in cellular phones. The idea of employing a heterojunction is as old as the conventional BJT, dating back to a patent from 1951.

Herbert Kroemer

Herbert Kroemer, a professor of electrical and computer engineering at the University of California, Santa Barbara, received his Ph.D. in theoretical physics in 1952 from the University of Göttingen, Germany, with a dissertation on hot electron effects in the then-new transistor, setting the stage for a career in research on the physics of semiconductor devices.

In 2000, Dr. Kroemer, along with Zhores I. Alferov, was awarded a Nobel Prize in Physics for developing semiconductor hetero-structures used in high-speed and opto-electronics. The other co-recipient of the Nobel Prize was Jack Kilby for his invention and development of integrated cicuits and micro-chips.

Born	25 August 1928 (age 81), Weimar, Germany
Nationality	Germany, United States
Fields	Electrical engineering, applied physics
Institutions	Fernmeldetechnisches Zentralamt, RCA Laboratories, Varian Associates, University of Colorado, University of California, Santa Barbara
Alma mater	University of Jena, University of Göttingen
Doctoral advisor	Fritz Sauter
Doctoral student	William Frensley
Influences	Friedrich Hund, Fritz Houtermans
Known for	Drift-field transistor, double-heterostructure laser
Notable awards	Nobel Prize in Physics (2000)

ENERGY BAND OFFSETS IN REAL HETEROJUNCTIONS (2000)

In real semiconductor heterojunctions, Anderson's model fails to predict actual band offsets. Some material parameters do not match experimental results using Anderson's rule. This idealized model ignores the fact that each material is made up of a crystal lattice whose electrical properties depend on a periodic arrangement of atoms. This periodicity is broken at the heterojunction interface to varying degrees. In cases where both materials have the same lattice, they may still have differing lattice constants which give rise to crystal strain which changes the band energies. In other cases the strain is relaxed via dislocations and other interfacial defects which also change the band energies.

A common anion rule was proposed, which guesses that since the valence band is related to anionic states, materials with the same anions should have very small valence band offsets. This however did not explain the data but is related to the trend that two materials with different anions tend to have larger valence band offsets than conduction band offsets.

Tersoff proposed a model based on more familiar metal-semiconductor junctions where the conduction band offset is given by the difference in Schottky barrier height. This model includes a dipole later at the interface between the two semiconductors which arises from electron tunneling from the conduction band of one material into the gap of the other. This model agrees well with systems where both materials are closely lattice matched such as GaAs/AlGaAs.

Anderson's rule overestimates the offset in the conduction band of the commercially important type two offset GaAs/AlAs system with $\Delta E_C / \Delta E_G = 0.73$ from the data above. It has been shown that for the actual ratio is closer $\Delta E_C / \Delta E_G = 0.6$. This is known as the 60 : 40 rule and applies to heterojunctions of GaAs with all compositions of AlGaAs. The typical method for measuring band offsets is by calculating them from measuring exciton energies in the luminescence spectra.

Jack St. Clair Kilby

Jack St. Clair Kilby was a Nobel laureate in physics in 2000 for his invention of the integrated circuit in 1958 while working at Texas Instruments (TI). He is also the inventor of the handheld calculator and thermal printer.

Born	8 November 1923, Jefferson City, Missouri, USA
Died	20 June 2005 (age 81), Dallas, Texas, USA
Nationality	United States
Fields	Physics, electrical engineering
Institutions	Texas instruments
Notable awards	Nobel Prize in Physics (2000), IEEE Medal of Honor

INTEGRATED CIRCUIT (2000)

In electronics, an integrated circuit (also known as IC, microcircuit, microchip, silicon chip, or chip) is a miniaturized electronic circuit (consisting mainly of semiconductor devices, as well as passive components) that has been manufactured in the surface of a thin substrate of semiconductor material. Integrated circuits are used in almost all electronic equipment in use today and have revolutionized the world of electronics. A hybrid integrated circuit is a miniaturized electronic circuit constructed of individual semiconductor devices, as well as passive components, bonded to a substrate or circuit board.

The idea of the integrated circuit was conceived by a radar scientist working for the Royal Radar Establishment of the British Ministry of Defence, Geoffrey W.A. Dummer (1909–2002), who published it at the Symposium on Progress in Quality Electronic Components in Washington, D.C. on 7 May 1952. He gave many symposia publicly to propagate his ideas. Dummer unsuccessfully attempted to build such a circuit in 1956.

Jack Kilby recorded his initial ideas concerning the integrated circuit in July 1958 and successfully demonstrated the first working integrated circuit on 12 September 1958. In his patent application of 6 February 1959, Kilby described his new device as a body of semiconductor material wherein all the components of the electronic circuit are completely integrated. Kilby won the 2000 Nobel Prize in Physics for his part of the invention of the integrated circuit.

Robert Noyce also came up with his own idea of an integrated circuit half a year later than Kilby. Noyce's chip solved many practical problems that Kilby's had not. Noyce's chip, made at Fairchild Semiconductor, was made of silicon, whereas Kilby's chip was made of germanium.

Early developments of the integrated circuit go back to 1949, when the German engineer Werner Jacobi (Siemens AG) filed a patent for an integrated-circuit-like semiconductor amplifying device showing five transistors on a common substrate arranged in a 2-stage amplifier arrangement. Jacobi discloses small and cheap hearing aids as typical industrial applications of his patent. A commercial use of his patent has not been reported.

A precursor idea to the IC was to create small ceramic squares (wafers), each one containing a single miniaturized component. Components could then be integrated and wired into a bidimensional or tridimensional compact grid. This idea, which looked very promising in 1957, was proposed to the US Army by Jack Kilby, and led to the short-lived Micromodule Program (similar to 1951s Project Tinkertoy). However, as the project was gaining momentum, Kilby came up with a new, revolutionary design: the IC.

Robert Noyce credited Kurt Lehovec of Sprague Electric for the *principle of p-n junction isolation* caused by the action of a biased p-n junction (the diode) as a key concept behind the IC.

Eric Allin Cornell

Eric Allin Cornell, is an American physicist who, along with Carl E. Wieman, was able to synthesize the first Bose-Einstein condensate in 1995. For their efforts, Cornell, Wieman, and Wolfgang Ketterle shared the Nobel Prize in Physics in 2001.

Born	19 December 1961 (age 48), Palo Alto, California, USA
Residence	Boulder, Colorado, USA
Nationality	United States
Fields	Physics
Institutions	University of Colorado, National Institute of Standards and Technology (NIST)
Alma mater	Massachusetts Institute of Technology (MIT), Stanford University
Known for	Bose-Einstein condensation
Notable awards	Lorentz Medal (1998), Nobel Prize in Physics (2001), Benjamin Franklin Medal in Physics (2000)

BOSE-EINSTEIN CONDENSATE (2001)

A Bose-Einstein condensate (BEC) is a state of matter of a dilute gas of weakly interacting bosons confined in an external potential and cooled to temperatures very near to absolute zero (0 K or −273.15°C). Under such conditions, a large fraction of the bosons occupy the lowest quantum state of the external potential, at which point quantum effects become apparent on a macroscopic scale.

This state of matter was first predicted by Satyendra Nath Bose and Albert Einstein in 1924–25. Bose first sent a paper to Einstein on the quantum statistics of light quanta (now called photons). Einstein was impressed, translated the paper himself from English to German and submitted it for Bose to the *Zeitschrift für Physik* which published it. Einstein then extended Bose's ideas to material particles (or matter) in two other papers.

Seventy years later, the first gaseous condensate was produced by Eric Cornell and Carl Wieman in 1995 at the University of Colorado at Boulder NIST-JILA lab, using a gas of rubidium atoms cooled to 170 nanokelvin (nK) (1.7×10^{-7} K). For their achievements Cornell, Wieman, and Wolfgang Ketterle at MIT received the 2001 Nobel Prize in Physics.

The slowing of atoms by use of cooling apparatus produces a singular quantum state known as a Bose condensate or Bose-Einstein condensate. This phenomenon was predicted in 1925 by generalizing Satyendra Nath Bose's work on the statistical mechanics of photons to atoms. The result of the efforts of Bose and Einstein is the concept of a Bose gas, governed by Bose-Einstein statistics, which describes the statistical distribution of identical particles with integer spin, now known as Bosons. Bosonic particles, which include the photon as well as atoms such as helium-4, are allowed to share quantum states with each other. Einstein demonstrated that cooling bosonic atoms to a very low temperature would cause them to fall into the lowest accessible quantum state, resulting in a new form of matter.

Consider a collection of N noninteracting particles, which can each be in one of two quantum states, $|0\rangle$ and $|1\rangle$. If the two states are equal in energy, each different configuration is equally likely.

If we can tell which particle is which, there are 2^N different configurations, since each particle can be in $|0\rangle$ or $|1\rangle$ independently. In almost all of the configurations, about half the particles are in $|0\rangle$ and the other half in $|1\rangle$. The balance is a statistical effect: the number of configurations is largest when the particles are divided equally.

Wolfgang Ketterle

Wolfgang Ketterle, is a German physicist and professor of physics at the Massachusetts Institute of Technology (MIT). His research has focused on experiments that trap and cool atoms to temperatures close to absolute zero, and he led one of the first groups to realize Bose-Einstein condensation in these systems in 1995.

For this achievement, as well as early fundamental studies of condensates, he was awarded the Nobel Prize in Physics in 2001, together with Eric Allin Cornell and Carl Wieman.

Born	21 October 1957 (age 52), Heidelberg, Germany
Nationality	Germany
Fields	Physics
Institutions	University of Heidelberg, MIT
Alma mater	Heidelberg, TUM, LMU, Max Planck Institute of Quantum Optics
Doctoral advisor	Herbert Walther, Hartmut Figger
Known for	Bose-Einstein condensates
Notable awards	The Benjamin Franklin Medal (2000), Nobel Prize in Physics (2001)

BOSE-EINSTEIN CONDENSATE (2001)

In 1938, Pyotr Kapitsa, John Allen and Don Misener discovered that helium-4 became a new kind of fluid, now known as a super fluid, at temperatures less than 2.17 K. Super fluid helium has many unusual properties, including zero viscosity and the existence of quantized vortices. It was quickly realized that the super fluidity was due to partial Bose Einstein condensation of the liquid. In fact, many of the properties of super fluid helium also appear in the gaseous Bose-Einstein condensates created by Cornell, Wieman and Ketterle. Super fluid helium-4 is a liquid rather than a gas, which means that the interactions between the atoms are relatively strong; the original theory of Bose-Einstein condensation must be heavily modified in order to describe it. Bose-Einstein condensation remains, however, fundamental to the super fluid properties of helium-4. Note that helium-3, consisting of fermions instead of bosons, also enters a super fluid phase at low temperature, which can be explained by the formation of bosonic Cooper pairs of two atoms each.

The first pure Bose-Einstein condensate was created by Eric Cornell, Carl Wieman, and co-workers at JILA on 5 June 1995. They did this by cooling a dilute vapor consisting of approximately two thousand rubidium-87 atoms to below 170 nK using a combination of laser cooling and magnetic evaporative cooling. About four months later, an independent effort led by Wolfgang Ketterle at MIT created a condensate made of sodium-23. Ketterle's condensate had about a hundred times more atoms, allowing him to obtain several important results such as the observation of quantum mechanical interference between two different condensates.

The Bose-Einstein condensation also applies to quasiparticles in solids. A magnon in an antiferromagnet carries spin 1 and thus obeys Bose-Einstein statistics. The density of magnons is controlled by an external magnetic field, which plays the role of the magnon chemical potential. This technique provides access to a wide range of boson densities from the limit of a dilute Bose gas to that of a strongly interacting Bose liquid. A magnetic ordering observed at the point of condensation is the analog of super fluidity. In 1999 Bose condensation of magnons was demonstrated in the antiferromagnet $TlCuCl_3$. The condensation was observed at temperatures as large as 14 K. Such a high transition temperature is due to the greater density achievable with magnons and the smaller mass. In 2006, condensation of magnons in ferromagnets was even shown at room temperature, where the authors used pumping techniques.

Carl Edwin Wieman

Carl Edwin Wieman is an American physicist at the University of British Columbia and Nobel Prize in Physics laureate in 2001 for his production of the first true Bose-Einstein condensate in 1995 with Eric Allin Cornell.

Born	26 March 1951 (age 59), Corvallis, Oregon
Nationality	United States
Fields	Physics
Institutions	University of British Columbia, University of Colorado
Alma mater	MIT, Stanford University
Known for	Bose-Einstein condensate
Notable awards	Lorentz Medal (1998), The Benjamin Franklin Medal (2000), Nobel Prize in Physics (2001), Oersted Medal (2007)

BOSE-EINSTEIN CONDENSATE (2001)

Further experimentation by the JILA team in 2000 uncovered a hitherto unknown property of Bose-Einstein condensates. Cornell, Wieman, and their coworkers originally used rubidium-87, an isotope whose atoms naturally repel each other, making a more stable condensate. The JILA team instrumentation now had better control over the condensate so experimentation was made on naturally *attracting* atoms of another rubidium isotope, rubidium-85 (having negative atom-atom scattering length). Through a process called Feshbach resonance involving a sweep of the magnetic field causing spin flip collisions, the JILA researchers lowered the characteristic, discrete energies at which the rubidium atoms bond into molecules, making their Rb-85 atoms repulsive and creating a stable condensate. The reversible flip from attraction to repulsion stems from quantum interference among condensate atoms which behave as waves.

When the scientists raised the magnetic field strength still further, the condensate suddenly reverted back to attraction, imploded and shrank beyond detection, and then exploded, expelling off about two-thirds of its 10,000 or so atoms. About half of the atoms in the condensate seemed to have disappeared from the experiment altogether, not being seen either in the cold remnant or the expanding gas cloud. Carl Wieman explained that under current atomic theory this characteristic of Bose-Einstein condensate could not be explained because the energy state of an atom near absolute zero should not be enough to cause an implosion; however, subsequent mean field theories have been proposed to explain it.

Because supernova explosions are also preceded by an implosion, the explosion of a collapsing Bose-Einstein condensate was named bosenova, a pun on the musical style bossa nova.

The atoms that seem to have disappeared almost certainly still exist in some form, just not in a form that could be accounted for in that experiment. Most likely they formed molecules consisting of two bor ded rubidium atoms. The energy gained by making this transition imparts a velocity sufficient for them to leave the trap without being detected.

Compared to more commonly encountered states of matter, Bose-Einstein condensates are extremely fragile. The slightest interaction with the outside world can be enough to warm them past the condensation threshold, eliminating their interesting properties and forming a normal gas. It is likely to be some time before any practical applications are developed.

Raymond Davis, Jr.

Raymond Davis, Jr. was an American chemist, physicist, and Nobel Prize winner in Physics. He shared the Nobel Prize in Physics in 2002 with Japanese physicist Masatoshi Koshiba and American Riccardo Giacconi for pioneering contributions to astrophysics, in particular for the detection of cosmic neutrinos, looking at the solar neutrino problem in the Homestake Experiment. He was almost 88 years old when awarded the prize, making him the oldest ever recipient of a Nobel Prize.

Born	14 October 1914, Washington, D.C., USA
Died	31 May 2006 (age 91), Blue Point, New York, USA
Nationality	United States
Fields	Chemist, physicist
Institutions	Monsanto Company, University of Pennsylvania
Alma mater	University of Maryland, College Park, Yale University
Known for	Neutrinos
Notable awards	Tom W. Bonner Prize (1988), Beatrice M. Tinsley Prize (1994) Wolf Prize in Physics (2000), National Medal of Science (2001) Nobel Prize in Physics (2002)

NEUTRINO (2002)

A neutrino (Meaning 'small neutral one') is an elementary particle that usually travels close to the speed of light, is electrically neutral, and is able to pass through ordinary matter almost undisturbed. This makes neutrinos extremely difficult to detect. Neutrinos have a very small, but nonzero mass. They are denoted by the Greek letter v (nu).

Neutrinos are created as a result of certain types of radioactive decay or nuclear reactions such as those that take place in the Sun, in nuclear reactors, or when cosmic rays hit atoms. There are three types, or flavours, of neutrinos: electron neutrinos, muon neutrinos and tau neutrinos; each type also has a corresponding antiparticle, called antineutrinos. Electron neutrinos (or antineutrinos) are generated whenever protons change into neutrons (or vice versa), the two forms of beta decay. Interactions involving neutrinos are mediated by the weak interaction.

Most neutrinos passing through the Earth emanate from the Sun, and more than 50 trillion solar neutrinos pass through the human body every second.

The neutrino was first postulated in 1930 by Wolfgang Pauli to preserve the conservation of energy, conservation of momentum, and conservation of angular momentum in beta decay the decay of an atomic nucleus (not known to contain or involve the neutron at the time) into a proton, an electron and an antineutrino.

$$n^0 \rightarrow p^+ + e^- + \nu^0$$

He theorized that an undetected particle was carrying away the observed difference between the energy, momentum, and angular momentum of the initial and final particles.

Pauli originally named his proposed light particle a *neutron*. When James Chadwick discovered a much more massive nuclear particle in 1932 and also named it a neutron, this left the two particles with the same name. Enrico Fermi, who developed the theory of beta decay, coined the term *neutrino* in 1934 as a clever way to resolve the confusion. It is the Italian equivalent of little neutral one.

Masatoshi Koshiba

 Masatoshi Koshiba, is a Japanese physicist who won the Nobel Prize in Physics in 2002. In the 1980s Koshiba, drawing on the work done by Raymond Davis Jr, constructed an underground neutrino detector in a zinc mine in Japan called Kamiokande II.

It was an enormous water tank surrounded by electronic detectors to sense flashes of light produced when neutrinos interacted with atomic nuclei in water molecules. Koshiba was able to confirm Davis's results that the Sun produces neutrinos and that fewer neutrinos were found than had been expected (a deficit that became known as the solar neutrino problem). In 1987 Kamiokande also detected neutrinos from a supernova explosion outside the Milky Way. After building a larger, more sensitive detector named Super-Kamiokande, which became operational in 1996, Koshiba found strong evidence for what scientists had already suspected that neutrinos, of which three types are known, change from one type into another in flight; this resolves the solar neutrino problem, since early experiments could only detect one type, not all three.

Born	19 September 1926 (age 83),Toyohashi, Aichi Prefecture, Japan
Nationality	Japan
Fields	Physics
Institutions	University of Chicago, University of Tokyo, Tokai University
Alma mater	University of Tokyo, University of Rochester
Known for	Astrophysics, Neutrinos
Notable awards	Nobel Prize in Physics (2002)

ASTROPHYSICS (2002)

Astrophysics (Greek: *Astro–* meaning 'star', and Greek: *physis –* meaning "nature") is the branch of astronomy that deals with the physics of the universe, including the physical properties of celestial objects such as galaxies, stars, planets, exoplanets, and the interstellar medium, as well as their interactions. The study of cosmology addresses questions of astrophysics at scales much larger than the size of particular gravitationally-bound objects in the universe.

Because astrophysics is a very broad subject, *astrophysicists* typically apply many disciplines of physics, including mechanics, electromagnetism, statistical mechanics, thermodynamics, quantum mechanics, relativity, nuclear and particle physics, and atomic and molecular physics. In practice, modern astronomical research involves a substantial amount of physics.

Although astronomy is as ancient as recorded history itself, it was long separated from the study of physics. In the Aristotelian worldview, the celestial world tended towards perfection bodies in the sky seemed to be perfect spheres moving in perfectly circular orbits while the earthly world seemed destined to imperfection; these two realms were not seen as related.

Aristarchus of Samos first put forward the notion that the motions of the celestial bodies could be explained by assuming that the Earth and all the other planets in the Solar System orbited the Sun. Unfortunately, in the geocentric world of the time, Aristarchus' heliocentric theory was deemed outlandish and heretical. For centuries, the apparently common-sense view that the Sun and other planets went round the Earth nearly went unquestioned until the development of Copernican heliocentrism in the 16th century AD. This was due to the dominance of the geocentric model developed by Ptolemy, a Hellenized astronomer from Roman Egypt, in his *Almagest* treatise.

Riccardo Giacconi

Riccardo Giacconi, is an Italian/American Nobel Prize-winning astrophysicist who laid the foundations of X-ray astronomy. Giacconi was awarded the Nobel Prize in Physics in 2002 for pioneering contributions to astrophysics, which have led to the discovery of cosmic X-ray sources.

He has simultaneously held the position of professor of physics and astronomy (1982–1997) and research professor (since 1998) at Johns Hopkins University, and is now University Professor. During this time, he was also Director General of the European Southern Observatory (ESO) (1993–1999). He is currently principal investigator for the Chandra Deep Field-South project with NASAs Chandra X-ray Observatory.

Born	6 October 1931 (age 78), Genoa, Italy
Residence	United States
Nationality	Italy, United States
Fields	Physics
Institutions	Johns Hopkins University, Chandra X-ray Observatory
Alma mater	University of Milan
Known for	Astrophysics
Notable awards	Nobel Prize in Physics (2002), Elliott Cresson Medal (1980)

ASTROPHYSICS (2002)

In the 9th century AD, the Persian physicist and astronomer, Ja'far Muhammad ibn Musaibn Shakir, hypothesized that the heavenly bodies and celestial spheres are subject to the same laws of physics as Earth, unlike the ancients who believed that the celestial spheres followed their own set of physical laws different from that of Earth. He also proposed that there is a force of attraction between heavenly bodies, vaguely foreshadowing the law of gravity.

In the early 11th century, the Arabic Ibn al-Haytham wrote the *Maqala fi daw al-qamar* (*On the Light of the Moon*) some time before 1021. This was the first successful attempt at combining mathematical astronomy with physics, and the earliest attempt at applying the experimental method to astronomy and astrophysics. He disproved the universally held opinion that the moon reflects sunlight like a mirror and correctly concluded that it emits light from those portions of its surface which the sun's light strikes. In order to prove that light is emitted from every point of the moon's illuminated surface, he built an ingenious experimental device. Ibn al-Haytham had formulated a clear conception of the relationship between an ideal mathematical model and the complex of observable phenomena; in particular, he was the first to make a systematic use of the method of varying the experimental conditions in a constant and uniform manner, in an experiment showing that the intensity of the light-spot formed by the projection of the moonlight through two small apertures onto a screen diminishes constantly as one of the apertures is gradually blocked up.

In the 14th century, Ibn al-Shatir produced the first model of lunar motion which matched physical observations, and which was later used by Copernicus. In the 13th to 15th centuries, Tusi and Ali Qushji provided the earliest empirical evidence for the Earth's rotation, using the phenomena of comets to refute Ptolemy's claim that a stationery Earth can be determined through observation. Kuscu further rejected Aristotelian physics and natural philosophy, allowing astronomy and physics to become empirical and mathematical instead of philosophical. In the early 16th century, the debate on the Earth's motion was continued by Al-Birjandi, who in his analysis of what might occur if the Earth were rotating, develops a hypothesis similar to Galileo Galilei's notion of circular inertia.

The small or large rock will fall to the Earth along the path of a line that is perpendicular to the plane (*sath*) of the horizon; this is witnessed by experience (*tajriba*). And this perpendicular is away from the tangent point of the Earth's sphere and the plane of the perceived (*hissi*) horizon. This point moves with the motion of the Earth and thus there will be no difference in place of fall of the two rocks.

Alexei Alexeyevich Abrikosov

 Alexei Alexeyevich Abrikosov, is a Russian theoretical physicist whose main contributions are in the field of condensed matter physics. He was awarded the Nobel Prize in Physics in 2003.

Born	25 June 1928 (age 81), Moscow, Russian SFSR, USSR
Nationality	Russian
Fields	Physics
Institutions	Landau Institute, Moscow State University Argonne National Laboratory
Alma mater	Moscow State University, USSR Academy of Sciences
Known for	Condensed matter physics
Notable awards	Nobel Prize in Physics (2003)
He is the son of the physician Alexei Ivanovich Abrikosov	

CONDENSED MATTER PHYSICS (2003)

Condensed matter physics is the field of physics that deals with the macroscopic and microscopic physical properties of matter. In particular, it is concerned with the condensed phases that appear whenever the number of constituents in a system is extremely large and the interactions between the constituents are strong. The most familiar examples of condensed phases are solids and liquids, which arise from the electromagnetic forces between atoms. More exotic condensed phases include the superconducting phase exhibited by certain materials at low temperature, the ferromagnetic and antiferromagnetic phases of spins on atomic lattices, and the Bose-Einstein condensate found in certain ultra cold atomic systems.

The aim of condensed matter physics is to understand the behavior of these phases by using well-established physical laws, in particular those of quantum mechanics, electromagnetism and statistical mechanics. The diversity of systems and phenomena available for study makes condensed matter physics by far the largest field of contemporary physics. By one estimate, one third of all United States physicists identify themselves as condensed matter physicists. The field has a large overlap with chemistry, materials science, and nanotechnology, and there are close connections with the related fields of atomic physics and biophysics. Theoretical condensed matter physics also shares many important concepts and techniques with theoretical particle and nuclear physics.

Historically, condensed matter physics grew out of solid-state physics, which is now considered one of its main subfields. The name of the field was apparently coined in 1967 by Philip Anderson and Volker Heine when they renamed their research group in the Cavendish Laboratory of the University of Cambridge from Solid-State Theory to Theory of Condensed Matter. In 1978, the Division of Solid State Physics at the American Physical Society was renamed as the Division of Condensed Matter Physics. One of the reasons for this change is that many of the concepts and techniques developed for studying solids can also be applied to fluid systems. For instance, the conduction electrons in an electrical conductor form a Fermi liquid, with similar properties to conventional liquids made up of atoms or molecules. Even the phenomenon of superconductivity, in which the quantum-mechanical properties of the electrons lead to collective behavior fundamentally different from that of a classical fluid, is closely related to the super fluid phase of liquid helium.

Vitaly Lazarevich Ginzburg

Vitaly Lazarevich Ginzburg was a Soviet theoretical physicist, astrophysicist, Nobel laureate, a member of the Russian Academy of Sciences and one of the fathers of Soviet hydrogen bomb. He was the successor to Igor Tamm as head of the Department of Theoretical Physics of the Academy's physics institute (FIAN), and an outspoken atheist.

Born	4 October 1916, Moscow, Russian Empire
Died	8 November 2009 (age 93), Moscow, Russia
Nationality	Russia
Fields	Theoretical physics
Institutions	P. N. Lebedev Physical Institute
Alma mater	Moscow State University
Doctoral advisor	Igor Tamm
Known for	Plasmas, superfluidity
Notable awards	Nobel Prize in Physics (2003), Wolf Prize in Physics (1994/95)

PLASMA (2003)

In physics and chemistry, plasma is a gas in which a certain portion of the particles are ionized. The presence of a non-negligible number of charge carriers makes the plasma electrically conductive so that it responds strongly to electromagnetic fields. Plasma, therefore, has properties quite unlike those of solids, liquids, or gases and is considered to be a distinct state of matter. Like gas, plasma does not have a definite shape or a definite volume unless enclosed in a container; unlike gas, in the influence of a magnetic field, it may form structures such as filaments, beams and double layers. Some common plasma's are stars and neon signs.

Plasma was first identified in a Crookes tube, and so described by Sir William Crookes in 1879 (he called it radiant matter). The nature of the Crookes tube cathode ray matter was subsequently identified by British physicist Sir J.J. Thomson in 1897, and dubbed plasma by Irving Langmuir in 1928, perhaps because it reminded him of a blood plasma.

Langmuir wrote: Except near the electrodes, where there are *sheaths* containing very few electrons, the ionized gas contains ions and electrons in about equal numbers so that the resultant space charge is very small. We shall use the name *plasma* to describe this region containing balanced charges of ions and electrons.

Plasmas are by far the most common phase of matter in the universe, both by mass and by volume. All the stars are made of plasma, and even the space between the stars is filled with plasma, albeit a very sparse one. In our solar system, the planet Jupiter accounts for most of the *non*-plasma, only about 0.1% of the mass and 10^{-15}% of the volume within the orbit of Pluto. Very small grains within gaseous plasma will also pick up a net negative charge, so that they in turn may act like a very heavy negative ion component of the plasma.

A dusty plasma is a plasma containing nanometer or micrometer-sized particles suspended in it. Dust particles may be charged and the plasma and particles behave as a plasma, following electromagnetic laws for particles up to about 10 nm (or 100 nm if large charges are present). Dust particles may form larger particles resulting in grain plasmas.

Dusty plasmas are encountered in Industrial processing plasmas and Space plasmas.

Sir Anthony James Leggett

Sir Anthony James Leggett, aka Tony Leggett, has been a Professor of Physics at the University of Illinois at Urbana-Champaign since 1983.

Born	26 March 1938 (age 72), Camberwell, London, England, UK
Residence	United States
Citizenship Dual	United Kingdom-United States
Fields	Physicist
Institutions	University of Sussex, University of Illinois at Urbana-Champaign
Alma mater	Oxford University
Doctoral advisor	Dirk ter Haar
Doctoral students	Amir O. Caldeira
Known for	Caldeira-Leggett model
Foundations of	quantum mechanics, superfluid phase of helium-3 quantum decoherence
Notable awards	Maxwell Medal and Prize (1975), Paul Dirac Medal (1992) Nobel Prize in Physics (2003), Wolf Prize in Physics (2002/03)

QUANTUM DISSIPATION (2003)

Quantum Dissipation is the branch of physics that studies the quantum analogues of the process of irreversible loss of energy observed at the classical level. Its main purpose is to derive the laws of classical dissipation from the framework of quantum mechanics. It shares many features with the subjects of quantum decoherence and quantum theory of measurement.

The main problem to address to study dissipation at the quantum level is the way to envisage the mechanism of irreversible loss of energy. Quantum mechanics usually deal with the Hamiltonian formalism, where the total energy of the system is a conserved quantity. So in principle it would not be possible to describe dissipation in this framework.

The idea to overcome this issue consists on splitting the total system in two parts: the quantum system where dissipation occurs and a so-called environment, or bath where the energy of the former will flow towards. The way both systems are coupled depends on the details of the microscopic model, and hence, the description of the bath. To include an irreversible flow of energy, requires that the bath contain an infinite number of degrees of freedom. Notice that by virtue of the principle of universality, it is expected that the particular description of the bath will not affect the essential features of the dissipative process, as far as the model contains the minimal ingredients to provide the effect.

QUANTUM DECOHERENCE

In quantum mechanics, quantum decoherence (also known as dephasing) is the mechanism by which quantum systems interact with their environments to exhibit probabilistically additive behavior. Quantum decoherence gives the *appearance* of wave function collapse and justifies the framework and intuition of classical physics as an acceptable approximation: decoherence is the mechanism by which the classical limit emerges out of a quantum starting point and it determines the location of the quantum-classical boundary. Decoherence occurs when a system interacts with its environment in a thermodynamically irreversible way. This prevents different elements in the quantum superposition of the system + environment's wave function from interfering with each other. Decoherence has been a subject of active research since the 1980s.

Decoherence can be viewed as the loss of information from a system into the environment (often modeled as a heat bath). Viewed in isolation, the system's dynamics are non-unitary (although the combined system plus environment evolves in a unitary fashion). Thus the dynamics of the system alone, treated in isolation from the environment, are irreversible. As with any coupling, entanglements are generated between the system and environment.

David Jonathan Gross

David Jonathan Gross, is an American particle physicist and string theorist. Along with Frank Wilczek and David Politzer, he was awarded the 2004 Nobel Prize in Physics for his discovery of asymptotic freedom. He is currently the director and holder of the Frederick W. Gluck Chair in Theoretical Physics at the Kavli Institute for Theoretical Physics of the University of California, Santa Barbara.

Born	19 February 1941 (age 69), Washington, D.C., USA
Nationality	United States
Residence	United States
Fields	Physicist
Institutions	University of California, Santa Barbara, Harvard University, Princeton University
Alma mater	Hebrew University, University of California, Berkeley
Doctoral advisor	Geoffrey Chew
Doctoral students	Frank Wilczek, Edward Witten, William E. Caswell, Rajesh Gopakumar
Known for	Asymptotic freedom, Heterotic string
Notable awards	Dirac Medal (1988), Nobel Prize in Physics (2004)

ASYMPTOTIC FREEDOM (2004)

In physics, asymptotic freedom is a property of some gauge theories that causes interactions between particles, such as quarks, to become arbitrarily weak at shorter distances, i.e. length scales that asymptotically converge to zero (or, equivalently, energy scales that become arbitrarily large).

Asymptotic freedom implies that in high-energy scattering the quarks move within nucleons, such as the neutron and proton, mostly as free non-interacting particles. It allows physicists to calculate the cross sections of various events in particle physics reliably using parton techniques.

Asymptotic freedom is a feature of quantum chromo dynamics (QCD), the quantum field theory of the interactions of quarks and gluons which was discovered in 1973 by David Gross and Frank Wilczek, and by David Politzer. Although these authors were the first to understand the physical relevance to the strong interactions, in 1969 Iosif Khriplovich discovered asymptotic freedom in the SU(2) gauge theory as a mathematical curiosity, and Gerardus 't Hooft in 1972 also noted the effect but did not publish. For their discovery, Gross, Wilczek and Politzer were awarded the Nobel Prize in Physics in 2004.

The discovery was instrumental in rehabilitating quantum field theory. Prior to 1973, many theorists suspected that field theory was fundamentally inconsistent because the interactions become infinitely strong at short-distances. This phenomenon is usually called a Landau pole, and it defines the smallest length scale that a theory can describe. This problem was discovered in field theories of interacting scalars and spinors, including quantum electrodynamics, and Lehman positivity led many to suspect that it is unavoidable. Asymptotically free theories become weak at short distances, there is no Landau pole, and these quantum field theories are believed to be completely consistent down to any length scale.

While the Standard Model is not entirely asymptotically free, in practice the Landau pole can only be a problem when thinking about the strong interactions. The other interactions are so weak that any inconsistency can only arise at distances shorter than the Planck length, where a field theory description is inadequate anyway.

Hugh David Politzer

Hugh David Politzer, is a theoretical physicist from the United States with Slovak ancestors. He shared the 2004 Nobel Prize in Physics with David Gross and Frank Wilczek for their discovery of asymptotic freedom in quantum chromodynamics.

Born	31 August 1949 (age 60), New York City, U.S.
Nationality	United States
Fields	Physics
Institutions	California Institute of Technology
Alma mater	University of Michigan, Harvard University
Doctoral advisor	Sidney Coleman
Known for	Quantum chromodynamics, asymptotic freedom
Notable awards	Nobel Prize in Physics (2004)

QUANTUM CHROMODYNAMICS (2004)

In theoretical physics, quantum chromodynamics (QCD) is a theory of the strong interaction (colour force), a fundamental force describing the interactions of the quarks and gluons making up hadrons (such as the proton, neutron or pion). It is the study of the SU(3) Yang-Mills theory of colour-charged fermions (the quarks). QCD is a quantum field theory of a special kind called a non-abelian gauge theory. It is an important part of the Standard Model of particle physics. A huge body of experimental evidence for QCD has been gathered over the years.

QCD enjoys two peculiar properties:

- Confinement, which means that the force between quarks does not diminish as they are separated. Because of this, it would take an infinite amount of energy to separate two quarks; they are forever bound into hadrons such as the proton and the neutron. Although analytically unproven, confinement is widely believed to be true because it explains the consistent failure of free quark searches, and it is easy to demonstrate in lattice QCD.

- Asymptotic freedom, which means that in very high-energy reactions, quarks and gluons interact very weakly. This prediction of QCD was first discovered in the early 1970s by David Politzer and by Frank Wilczek and David Gross. For this work they were awarded the 2004 Nobel Prize in Physics.

There is no phase-transition line separating these two properties: confinement in dominant in low-energy scales but, as energy increases, asymptotic freedom becomes dominant.

The word *quark* was coined by American physicist Murray Gell-Mann in its present sense, the word having been taken from the phrase Three quarks for Muster Mark in *Finnegans Wake* by James Joyce. Gell-Mann wrote in a private letter of 27 June 1978, to the editor of the Oxford English Dictionary that he had been influenced by Joyce's words: The allusion to three quarks seemed perfect (originally there were only three subatomic quarks.) Gell-Mann, however, wanted to pronounce the word with (ô) not (ä), as Joyce seemed to indicate by rhyming words in the vicinity such as *Mark*. Gell-Mann got around that by supposing that one ingredient of the line three quarks for Muster Mark was a cry of Three quarts for Mister heard in H.C. Earwicker's pub, a plausible suggestion given the complex punning in Joyce's novel.

The three kinds of charge in QCD (as opposed to one in quantum electrodynamics or QED) are usually referred to as colour charge by loose analogy to the three kinds of colour (red, green and blue) perceived by humans. Other than this clever nomenclature, the quantum parameter colour is completely unrelated to the everyday, familiar phenomenon of colour.

Since the theory of electric charge is dubbed electrodynamics, the Greek word chroma (meaning colour) is applied to the theory of colour charge, chromodynamics.

Frank Anthony Wilczek

Frank Anthony Wilczek, is a theoretical physicist from the United States and a Nobel laureate. He is currently the Herman Feshbach Professor of Physics at the Massachusetts Institute of Technology. Wilczek, along with David Gross and H. David Politzer, was awarded the Nobel Prize in Physics in 2004 for their discovery of asymptotic freedom in the theory of the strong interaction.

Born	15 May 1951 (age 59), Mineola, New York, U.S.
Nationality	United States
Fields	Physics
Institutions	MIT
Alma mater	University of Chicago, Princeton University
Doctoral advisor	David Gross
Doctoral students	Mark Alford, Michael Forbes, Martin Greiter, Christoph Holzhey, David Kessler, Finn Larsen Richard MacKenzie, John March-Russell Chetan Nayak, Maulik Parikh, Krishna Rajagopal David Robertson, Sean Robinson, Alfred Shapere Stephen Wandzura
Known for	Quantum chromodynamics
Notable awards	Lorentz Medal (2002), Nobel Prize in Physics (2004)

QUANTUM CHROMODYNAMICS (2004)

With the invention of bubble chambers and spark chambers in the 1950s, experimental particle physics discovered a large and ever-growing number of particles called hadrons. It seemed that such a large number of particles could not all be fundamental. First, the particles were classified by charge and isospin by Eugene Wigner and Werner Heisenberg; then, in 1953, according to strangeness by Murray Gell-Mann and Kazuhiko Nishijima. To gain greater insight, the hadrons were sorted into groups having similar properties and masses using the *eightfold way*, invented in 1961 by Gell-Mann and Yuval Ne'eman. Gell-Mann and George Zweig, correcting an earlier approach of Shoichi Sakata, went on to propose in 1963, that the structure of the groups could be explained by the existence of three flavours of smaller particles inside the hadrons: the quarks.

Perhaps the first remark that quarks should possess an additional quantum number was made as a short footnote in the preprint of Boris Struminsky in connection with Ω^- hyperon composed of three strange quarks with parallel spins.

Boris Struminsky was a PhD student of Nikolay Bogolyubov. The problem considered in this preprint was suggested by Nikolay Bogolyubov, who advised Boris Struminsky in this research. In the beginning of 1965, Nikolay Bogolyubov, Boris Struminsky and Albert Tavchelidze wrote a preprint with a more detailed discussion of the additional quark quantum degree of freedom. This work was also presented by Albert Tavchelidze without obtaining consent of his collaborators for doing so at an international conference in Trieste (Italy), in May 1965.

A similar mysterious situation was with the Δ^{++} baryon; in the quark model, it is composed of three up quarks with parallel spins. In 1965, Moo-Young Han with Yoichiro Nambu and Oscar W. Greenberg independently resolved the problem by proposing that quarks possess an additional SU(3) gauge degree of freedom, later called colour charge. Han and Nambu noted that quarks might interact via an octet of vector gauge bosons: the gluons.

Since free quark searches consistently failed to turn up any evidence for the new particles, and because an elementary particle back then was *defined* as a particle which could be separated and isolated, Gell-Mann often said that quarks were merely convenient mathematical constructs, not real particles. The meaning of this statement was usually clear in context: He meant quarks are confined, but he also was implying that the strong interactions could probably not be fully described by quantum field theory.

Roy Jay Glauber

Roy Jay Glauber, is an American theoretical physicist. He is the Mallinckrodt Professor of Physics at Harvard University and Adjunct Professor of Optical Sciences at the University of Arizona.

Born in New York City, he was awarded one half of the 2005 Nobel Prize in Physics for his contribution to the quantum theory of optical coherence, with the other half shared by John L. Hall and Theodor W. Hänsch. In this work, published in 1963, he created a model for photodetection and explained the fundamental characteristics of different types of light, such as laser light and light from light bulbs. His theories are widely used in the field of quantum optics.

Born	1 Sept. 1925 (age 84), New York City, New York, USA
Residence	United States
Nationality	United States
Fields	Physics
Institutions	Harvard University
Alma mater	Harvard University
Doctoral advisor	Julian Schwinger
Doctoral students	Daniel Frank Walls
Known for	Photodetection, quantum optics
Notable awards	Michelson Medal (1985), Nobel Prize in Physics (2005)

COHERENCE (2005)

In physics, coherence is a property of waves that enables stationary interference. More generally, coherence describes all properties of the correlation between physical quantities of a wave.

When interfering, two waves can add together to create a larger wave or subtract from each other to create a smaller wave depending on their relative phase. Two waves are said to be coherent if they have a constant relative phase. The degree of coherence is measured by the interference visibility, a measure of how perfectly the waves can cancel due to destructive interference.

Coherence was originally introduced in connection with Young's double-slit experiment in optics but is now used in any field that involves waves, such as acoustics, electrical engineering, neuroscience, and quantum physics. The property of coherence is the basis for commercial applications such as holography, the Sagnac gyroscope, radio antenna arrays, optical coherence tomography and telescope interferometers.

The coherence of two waves follows from how well correlated the waves are as quantified by the cross-correlation function. The cross-correlation quantifies the ability to predict the value of the second wave by knowing the value of the first. As an example, consider two waves perfectly correlated for all times. At any time, if the first wave changes, the second will change in the same way. If combined they can exhibit complete constructive interference/superposition at all times. It follows that they are perfectly coherent. As will be discussed below, the second wave need not be a separate entity. It could be the first wave at a different time or position. In this case, the measure of correlation is the autocorrelation function.

These states are unified by the fact that their behavior is described by a wave equation or some generalization thereof. Waves in a rope, Surface waves in a liquid, Electric signals (fields) in transmission cables, Sound, Radio waves and Microwaves, Light waves (optics), Electrons, atoms and any other object (such as a baseball, as described by quantum physics). In most of these systems, one can measure the wave directly. Consequently, its correlation with another wave can simply be calculated. However, in optics one cannot measure the electric field directly as it oscillates much faster than any detector's time resolution. Instead, we measure the intensity of the light. Most of the concepts involving coherence which will be introduced below were developed in the field of optics and then used in other fields. Therefore, many of the standard measurements of coherence are indirect measurements, even in fields where the wave can be measured directly.

John Lewis

John Lewis "Jan" Hall, is an American physicist, and Nobel laureate in physics. He shared one half of the 2005 Nobel Prize in Physics with Theodor W. Hänsch for his work in precision spectroscopy.

Born	21 August 1934 (age 75), Denver, Colorado, U.S.
Nationality	United States
Fields	Physics
Institutions	University of Colorado, JILA, NIST
Alma mater	Carnegie Institute of Technology
Known for	Optical frequency comb
Notable awards	Nobel Prize in Physics (2005)

FREQUENCY COMB (2005)

A frequency comb is the graphic representation of the spectrum of a mode locked laser. An octave spanning comb can be used for mapping radio frequencies into the optical frequency range or it can be used to steer a piezoelectric mirror within a carrier envelope phase correcting feedback loop. It should not to be confused with mono-mode laser frequency stabilization as mode-locking requires multi-mode lasers.

Modelocked lasers produce a series of optical pulses separated in time by the round-trip time of the laser cavity. The spectrum of such a pulse train is a series of Dirac delta functions separated by the repetition rate of the laser. This series of sharp spectral lines is called a frequency comb.

A purely electronic device, which generates a series of pulses, also generates a frequency comb. These are produced for electronic sampling oscilloscopes, but also used for frequency comparison of microwaves, because they reach up to 1 THz. Since they include 0 Hz they do not need the tricks which make up the rest of this article.

This requires broadening of the laser spectrum so that it spans an octave. This is usually done using highly nonlinear photonic crystal fiber. However, it has been shown that an octave-spanning spectrum can be generated directly from a Ti: sapphire laser using intracavity self-phase modulation. Or the second harmonic can be generated in a long crystal so that by consecutive sum frequency generation and difference frequency generation the spectrum of first and second harmonic widens until they overlap. Broadening to an octave is typically achieved using super continuum generation by strong self-phase modulation in nonlinear photonic crystal fiber.

Each line is displaced from a harmonic of the repetition rate by the carrier-envelope offset frequency. The carrier-envelope offset frequency is the rate at which the peak of the carrier frequency slips from the peak of the pulse envelope on a pulse-to-pulse basis.

Measurement of the carrier-envelope offset frequency is usually done with a self-referencing technique, in which the phase of one part of the spectrum is compared to its harmonic.

In the frequency $-2\times$ frequency technique, light at the lower energy side of the broadened spectrum is doubled using second harmonic generation in a nonlinear crystal and a heterodyne beat is generated between that and light at the same wavelength on the upper energy side of the spectrum. This beat frequency, detectable with a photodiode, is the carrier-envelope offset frequency.

Alternatively, from light at the higher energy side of the broadened spectrum the frequency at the peak of the spectrum is subtracted in a nonlinear crystal and a heterodyne beat is generated between that and light at the same wavelength on the lower energy side of the spectrum. This beat frequency, detectable with a photodiode, is the carrier-envelope offset frequency.

Theodor Wolfgang Hänsch

 Theodor Wolfgang Hänsch, is a German physicist. He received the 2005 Nobel Prize in Physics for contributions to the development of laser-based precision spectroscopy, including the optical frequency comb technique, sharing the prize with John L. Hall and Roy J. Glauber.

Born	30 October 1941 (age 68), Heidelberg, Germany
Nationality	Germany
Fields	Physics
Institutions	Ludwig-Maximilians University, Max-Planck-Institute Stanford University
Alma mater	University of Heidelberg
Doctoral students	Carl E. Wieman, Markus Greiner
Known for	Laser-based precision spectroscopy
Notable awards	Nobel Prize in Physics (2005) Gottfried Wilhelm Leibniz Prize(1989)

NATURE OF EXCITATION MEASURED (2005)

The type of spectroscopy depends on the physical quantity measured. Normally, the quantity that is measured is an intensity, of energy either absorbed or produced.

- Electromagnetic spectroscopy involves interactions of matter with electromagnetic radiation, such as light.
- Electron spectroscopy involves interactions with electron beams. Auger spectroscopy involves inducing the Auger effect with an electron beam. In this case the measurement typically involves the kinetic energy of the electron as variable.
- Acoustic spectroscopy involves the frequency of sound.
- Dielectric spectroscopy involves the frequency of an external electrical field.

- Mechanical spectroscopy involves the frequency of an external mechanical stress, e.g. a torsion applied to a piece of material.

MEASUREMENT PROCESS

Most spectroscopic methods are differentiated as either atomic or molecular based on whether or not they apply to atoms or molecules. Along with that distinction, they can be classified on the nature of their interaction:

- Absorption spectroscopy uses the range of the electromagnetic spectra in which a substance absorbs. This includes atomic absorption spectroscopy and various molecular techniques, such as infrared, ultraviolet-visible and microwave spectroscopy.
- Emission spectroscopy uses the range of electromagnetic spectra in which a substance radiates (emits). The substance first must absorb energy. This energy can be from a variety of sources, which determines the name of the subsequent emission, like luminescence. Molecular luminescence techniques include spectrofluorimetry.
- Scattering spectroscopy measures the amount of light that a substance scatters at certain wavelengths, incident angles, and polarization angles. One of the most useful applications of light scattering spectroscopy is Raman spectroscopy.
- Spectroscopy is the study of the interaction between matter and radiated energy. Historically, spectroscopy originated through the study of visible light dispersed according to its wavelength, e.g. by a prism. Later the concept was expanded greatly to comprise any interaction with radiative energy as a function of its wavelength or frequency. Spectroscopic data is often represented by a spectrum, a plot of the response of interest as a function of wavelength or frequency.

John Cromwell Mather

 John Cromwell Mather, is an American astrophysicist, cosmologist and Nobel Prize in Physics laureate for his work on COBE with George Smoot. Its goals were to investigate the cosmic microwave background radiation (CMB) of the universe and provide measurements that would help shape our understanding of the cosmos.

This work helped cement the Big Bang theory of the universe using the Cosmic Background Explorer Satellite (COBE). According to the Nobel Prize committee, the COBE-project can also be regarded as the starting point for cosmology as a precision science.

Born	7 August 1946 (age 63), Roanoke, Virginia, USA
Residence	United States
Nationality	United States
Fields	Astrophysics, cosmology
Institutions	NASA
Alma mater	Swarthmore College, University of California, Berkeley
Known for	Cosmic microwave background radiation
Notable awards	Nobel Prize in Physics (2006)

COSMIC MICROWAVE BACKGROUND RADIATION (2006)

In cosmology, cosmic microwave background (CMB) radiation (also CMBR, CBR, MBR, and relic radiation) is a form of electromagnetic radiation filling the universe. With a traditional optical telescope, the space between stars and galaxies (the *background*) is pitch black. But with a radiotelescope, there is a faint background glow, almost exactly the same in all directions, that is not associated with any star, galaxy, or other object. This glow is strongest in the microwave region of the radio spectrum, hence the name *cosmic microwave background radiation*.

The CMBR is well explained as radiation left over from an early stage in the creation of the universe, and its discovery is considered a landmark confirmation of the Big Bang model of the universe. When the universe was young, before the formation of stars and planets, it was smaller, much hotter, and filled with a uniform glow from its white-hot fog of hydrogen plasma. As the universe expanded, both the plasma and the radiation filling it, grew cooler. When the universe cooled enough, stable atoms could form. These atoms could no longer absorb the thermal radiation, and the universe became transparent instead of being an opaque fog. The photons that existed at that time have been propagating ever since, though growing fainter and less energetic, since the exact same photons fill a larger and larger universe. This is the source for the term *relic radiation*, another name for the CMBR.

Precise measurements of cosmic background radiation are critical to cosmology, since any proposed model of the universe must explain this radiation. The CMBR has a thermal black body spectrum at a temperature of 2.725 K, thus the spectrum peaks in the microwave range frequency of 160.2 GHz, corresponding to a 1.9 mm wavelength. The glow is almost but not quite uniform in all directions, and shows a very specific pattern equal to that expected if the inherent randomness of a red-hot gas is blown up to the size of the universe. In particular, the spatial power spectrum contains small anisotropies, or irregularities, which vary with the size of the region examined. They have been measured in detail, and match what would be expected if small thermal variations, generated by quantum fluctuations of matter in a very tiny space, had expanded to the size of the observable universe we can detect today. This is still a very active field of study, with scientists seeking both better data and better interpretations of the initial conditions of expansion.

George Fitzgerald Smoot III

George Fitzgerald Smoot III, is an American astrophysicist, cosmologist, Nobel laureate. He won the Nobel Prize in Physics in 2006 for his work on COBE with John C. Mather that led to the measurement of the black body form and anisotropy of the cosmic microwave background radiation.

Born	20 February 1945 (age 65), Yukon, Florida, U.S.
Residence	France
Nationality	United States
Fields	Physics
Institutions	UC Berkeley/LBNL/Université Paris Diderot-Paris 7
Alma mater	Massachusetts Institute of Technology
Doctoral advisor	David H. Frisch
Known for	Cosmic microwave background radiation
Notable awards	Albert Einstein Medal (2003), Nobel Prize in Physics (2006) Oersted Medal (2009)

COSMIC MICROWAVE BACKGROUND RADIATION (2006)

The cosmic microwave background is isotropic to roughly one part in 100,000: the root mean square variations are only 18 μK. The Far-Infrared Absolute Spectrophotometer (FIRAS) instrument on the NASA Cosmic Background Explorer (COBE) satellite has carefully measured the spectrum of the cosmic microwave background. The FIRAS project members compared the CMB with an internal reference black body and the spectra agreed to within the experimental error. They concluded that any deviations from the black body form that might still remain undetected in the CMB spectrum over the wavelength range from 0.5 to 5 mm must have a weighted rms value of at most 50 parts per million (0.005%) of the CMB peak brightness. This made the CMB spectrum the most precisely measured black body spectrum in nature.

The cosmic microwave background is perhaps the main prediction of the Big Bang model. In addition, Inflationary Cosmology predicts that after about 10^{-37} seconds the nascent universe underwent exponential growth that smoothed out nearly all in homogeneities. This was followed by symmetry breaking; a type of phase transition that set the fundamental forces and elementary particles in their present form. After 10^{-6} seconds, the early universe was made up of a hot plasma of photons, electrons, and baryons. The photons were constantly interacting with the plasma through Thomson scattering. As the universe expanded, adiabatic cooling caused the plasma to cool until it became favorable for electrons to combine with protons and form hydrogen atoms. This recombination event happened at around 3000 K or when the universe was approximately 379,000 years old. At this point, the photons scattered off the now electrically neutral atoms and began to travel freely through space, resulting in the decoupling of matter and radiation.

The colour temperature of the photons has continued to diminish ever since; now down to 2.725 K, their temperature will continue to drop as the universe expands. According to the Big Bang model, the radiation from the sky, we measure, today comes from a spherical surface called the surface of last scattering. This represents the collection of spots in space at which the decoupling event is believed to have occurred, less than 400,000 years after the Big Bang, and at a point in time such that the photons from that distance have just reached observers. The estimated age of the Universe is 13.75 billion years. However, because the Universe has continued expanding since that time, the moving distance from the Earth to the edge of the observable universe is now at least 46.5 billion light years.

The Big Bang theory suggests that the cosmic microwave background fills all of observable space, and that most of the radiation energy in the universe is in the cosmic microwave background, which makes up a fraction of roughly 6×10^{-5} of the total density of the universe.

Albert Fert

 Albert Fert, is a French physicist and one of the discoverers of giant magnetoresistance which brought about a breakthrough in gigabyte hard disks.

He is currently a professor at Université Paris-Sud in Orsay and scientific director of a joint laboratory ('Unité mixte de recherche') between the Centre national de la recherche scientifique (National Scientific Research Centre) and Thales Group. Also, he is an Adjunct professor of physics at Michigan State University. He was awarded the 2007 Nobel Prize in Physics together with Peter Grünberg.

Born	7 March 1938 (age 72), Carcassonne, France
Residence	Paris, France
Nationality	France
Fields	Physics
Institutions	Université Paris-Sud, Michigan State University
Alma mater	École normale supérieure
Doctoral advisor	I. A. Campbell
Known for	Giant magnetoresistive effect
Notable awards	Wolf Prize in Physics (2006) , Japan Prize (2007), Nobel Prize in Physics (2007)

GIANT MAGNETO RESISTANCE (2007)

Giant Magneto Resistance (GMR) is a quantum mechanical magneto resistance effect observed in thin film structures composed of alternating ferromagnetic and non magnetic layers. The 2007 Nobel Prize in physics was awarded to Albert Fert and Peter Grünberg for the discovery of GMR.

The effect is observed as a significant change in the electrical resistance depending on whether the magnetization of adjacent ferromagnetic layers are in a parallel or an antiparallel alignment. The overall resistance is relatively low for parallel alignment and relatively high for antiparallel alignment. GMR is used commercially by hard disk drive manufacturers.

GMR was first discovered in 1988, in Fe/Cr/Fe trilayers by a research team led by Peter Grünberg of the Jülich Research Centre (DE), who owns the patent. It was also simultaneously but independently discovered in Fe/Cr multilayers by the group of Albert Fert of the University of Paris-Sud (FR). The Fert group first saw the large effect in multilayers that led to its naming, and first correctly explained the underlying physics. The discovery of GMR is considered the birth of spintronics. Grünberg and Fert have received a number of prestigious prizes and awards for their discovery and contributions to the field of spintronics including the 2007 Nobel Prize in Physics.

Multilayer GMR:

In multilayer GMR two or more ferromagnetic layers are separated by a very thin (about 1 nm) non-ferromagnetic spacer (e.g. Fe/Cr/Fe). At certain thicknesses the RKKY coupling between adjacent ferromagnetic layers becomes antiferromagnetic, making it energetically preferable for the magnetizations of adjacent layers to align in anti-parallel. The electrical resistance of the device is normally higher in the anti-parallel case and the difference can reach more than 10% at room temperature. The interlayer spacing in these devices typically corresponds to the second antiferromagnetic peak in the AFM-FM oscillation in the RKKY coupling. The GMR effect was first observed in the multilayer configuration, with much early research into GMR focusing on multilayer stacks of 10 or more layers.

Peter Andreas Grünberg

Peter Andreas Grünberg, is a German physicist, and Nobel Prize winner in Physics laureate for his discovery with Albert Fert of giant magnetoresistance which brought about a breakthrough in gigabyte hard disk drives.

Born	18 May 1939 (age 71), Pilsen, Protectorate of Bohemia and Moravia
Nationality	Germany
Fields	Physics
Institutions	Carleton University, Jülich Research Centre, University of Cologne
Alma mater	Darmstadt University of Technology
Doctoral advisor	Stefan Hüfner
Known for	Giant magneto resistive effect
Notable awards	Wolf Prize in Physics (2006), European Inventor of the Year (2006) Japan Prize 2007, Nobel Prize in Physics (2007)

PSEUDO-SPIN VALVE (2007)

Pseudo-spin valve devices are very similar to the spin valve structures. The significant difference is the coercivities of the ferromagnetic layers. In a pseudo-spin valve structure a soft magnet will be used for one layer; where as a hard ferromagnet will be used for the other. This allows an applied field to flip the magnetization of the hard ferromagnet layer. For pseudo-spin valves, the non-magnetic layer thickness must be great enough so that exchange coupling minimized. This reduces the chance that the alignment of the magnetization of adjacent layers will spontaneously change at a later time.

GRANULAR GMR

Granular GMR is an effect that occurs in solid precipitates of a magnetic material in a non-magnetic matrix. Till date, granular GMR has only been observed in matrices of copper containing cobalt granules. The reason for this is that copper and cobalt are immiscible, and so it is possible to create the solid precipitate by rapidly cooling a molten mixture of copper and cobalt. Granule sizes vary depending on the cooling rate and amount of subsequent annealing. Granular GMR materials have not been able to produce the high GMR ratios found in the multilayer counterparts.

GMR AND TUNNEL MAGNETO RESISTANCE (TMR)

Tunnel magneto resistance (TMR) is an extension of spin valve GMR in which the electrons travel with their spins oriented perpendicularly to the layers across a thin insulating tunnel barrier (replacing the non ferromagnetic spacer). This allows larger impedance, a larger magneto resistance value (~10 × at room temperature) and a ~0 temperature coefficient to be achieved simultaneously. TMR has now replaced GMR in disk drives, in particular for high area densities and perpendicular recording. TMR has led to the emergence of MRAM memories and reprogrammable magnetic logic devices.

Applications: GMR has triggered the rise of a new field of electronics called spintronics which has been used extensively in the read heads of modern hard drives and magnetic sensors. A hard disk storing binary information can use the difference in resistance between parallel and ant parallel layer alignments as a method of storing 1s and 0s.

A high GMR is preferred for optimal data storage density. Current perpendicular-to-plane (CPP) Spin valve GMR currently yields the highest GMR. Research continues with older current-in-plane configuration and in the tunneling magneto resistance (TMR) spin valves which enable disk drive densities exceeding 1 Terabyte per square inch.

Yoichiro Nambu

Yoichiro Nambu, is a Japanese born American physicist, currently a professor at the University of Chicago. Known for his contributions to the field of theoretical physics, he was awarded the Nobel Prize in Physics in 2008 for the discovery of the mechanism of spontaneous broken symmetry in subatomic physics.

Born	18 January 1921 (age 89), Tokyo, Japan
Fields	Physics
Institutions	University of Tokyo (1942–49) Osaka City University (1949-52) Institute for Advanced Study (1952–54) University of Chicago (1954)
Known for	Spontaneous symmetry breaking
Notable awards	US National Medal of Science (1982), Dirac Medal (1986) J.J. Sakurai Prize (1994), Wolf Prize in Physics (1994/1995) Nobel Prize in Physics (2008)

SPONTANEOUS SYMMETRY BREAKING (2008)

Spontaneous symmetry breaking is the process by which a system described in a theoretically symmetrical way ends up in a non symmetrical state. Though the process by itself is interesting from a mathematical point of view, it is fairly simple. Its fame outside the scientific community stems from its use in the standard model of particle physics, one of the most fundamental theories of science. In the context of its use within the standard model, it is far more complicated.

For spontaneous symmetry breaking to occur, there must be a system in which there are several equally likely outcomes. The system as a whole is therefore symmetric with respect to these outcomes. However, if the system is sampled a specific outcome must occur. Though we know the system as a whole is symmetric, we also know that it is never encountered with this symmetry, only in one specific state. Because one of the outcomes is always found with probability 1, and the others with probability 0, they are no longer symmetric. Hence, the symmetry is said to be spontaneously broken in that theory.

When a theory is symmetric with respect to a symmetry group, but asserts that one element of the group is distinct, then spontaneous symmetry breaking has occurred. To be clear: the theory must not say which member is distinct, only that one is. From this point on, the theory can be treated as if this element actually is distinct, with the proviso that any results found in this way must be re symmetrised, by taking the average of each of the elements of the group being distinct one.

A common example to help explain this phenomenon is a ball sitting on top of a hill. This ball is in a completely symmetric state. However, its state is unstable: the slightest perturbing force will cause the ball to roll down the hill in some particular direction. At that point, symmetry has been broken because the direction in which the ball rolled has a feature that distinguishes it from all other directions.

Before spontaneous symmetry breaking, the Standard Model predicts the existence of all the required particles. However, some particles are predicted to be mass less, when in reality they have mass. Obviously, this is a major failing of the theory in that state. To overcome this, the Higgs mechanism uses spontaneous symmetry breaking to give these particle masses. It also predicts a new, as yet undetected particle, the Higgs boson. This particle is frequently mentioned within the media, as major experiments, such as those at CERN, are currently trying to find it. If the Higgs boson is not found, it will mean the Higgs mechanism and spontaneous symmetry breaking as they are currently used cannot be correct, and physicists must come up with a new model to explain the fundamental laws of nature. A more detailed presentation of this mechanism is given in the article on the Yukawa interaction, where it is shown how spontaneous symmetry breaking can be used to give mass to fermions.

Makoto Kobayashi

Makoto Kobayashi, is a Japanese physicist known for his work on CP-violation who was awarded one quarter of the 2008 Nobel Prize in Physics for the discovery of the origin of the broken symmetry which predicts the existence of at least three families of quarks in nature.

Born	7 April 1944 (age 66) , Nagoya, Japan
Citizenship	Japan
Fields	High energy physics (theory)
Institutions	Kyoto University, High Energy Accelerator Research Organization
Alma mater	Nagoya University
Doctoral advisor	Shoichi Sakata
Known for	Work on CP violation, CKM matrix
Notable awards	Sakurai Prize (1985), Japan Academy Prize (1985), Asahi Prize (1995), High Energy and Particle Physics Prize by European Physical Society (2007), Nobel Prize in Physics (2008)

CP VIOLATION (2008)

In particle physics, CP violation is a violation of the postulated CP symmetry: the combination of C symmetry and P symmetry. CP symmetry states that the laws of physics should be the same if a particle were interchanged with its antiparticle, and left and right were swapped. The discovery of CP violation in 1964 in the decays of neutral kaons resulted in the Nobel Prize in Physics in 1980 for its discoverers James Cronin and Val Fitch.

It plays an important role both in the attempts of cosmology to explain the dominance of matter over antimatter in the present Universe, and in the study of weak interactions in particle physics.

***CP* is the product of two symmetries:** C for charge conjugation, which transforms a particle into its antiparticle, and P for parity, which creates the mirror image of a physical system. The strong interaction and electromagnetic interaction seem to be invariant under the combined CP transformation operation, but this symmetry is slightly violated during certain types of weak decay. Historically, CP symmetry was proposed to restore order after the discovery of parity violation in the 1950s.

The idea behind parity symmetry is that the equations of particle physics are invariant under mirror inversion. This leads to the prediction that the mirror image of a reaction occurs at the same rate as the original reaction. Parity symmetry appears to be valid for all reactions involving electromagnetism and strong interactions. Until 1956, parity conservation was believed to be one of the fundamental geometric conservation laws. However, in 1956 a careful critical review of the existing experimental data by theoretical physicists Tsung-Dao Lee and Chen Ning Yang revealed that while parity conservation had been verified in decays by the strong or electromagnetic interactions, it was untested in the weak interaction. They proposed several possible direct experimental tests. The first test based on beta decay of Cobalt-60 nuclei was carried out in 1956 by a group led by Chien-Shiung Wu, and demonstrated conclusively that weak interactions violate the P symmetry or, as the analogy goes, some reactions did not occur as often as their mirror image.

Overall, the symmetry of a quantum mechanical system can be restored if another symmetry S can be found such that the combined symmetry PS remains unbroken. This rather subtle point about the structure of Hilbert space was realized shortly after the discovery of P violation, and it was proposed that charge conjugation was the desired symmetry to restore order.

Toshihide Maskawa or Masukawa

 Toshihide Maskawa or Masukawa, is a Japanese theoretical physicist known for his work on CP-violation who was awarded one quarter of the 2008 Nobel Prize in Physics for the discovery of the origin of the broken symmetry which predicts the existence of at least three families of quarks in nature.

Born	7 February 1940 (age 70), Nagoya, Japan
Residence	Japan
Nationality	Japanese
Fields	High energy physics (theory)
Institutions	Nagoya University, Kyoto University, Kyoto Sangyo University
Alma mater	Nagoya University
Doctoral advisor	Shoichi Sakata
Known for	Work on CP-violation, CKM matrix
Notable awards	Sakurai Prize (1985), Japan Academy Prize (1985), Asahi Prize (1994) Nobel Prize in Physics (2008)

CABIBBO-KOBAYASHI-MASKAWA MATRIX (2008)

In the Standard Model of particle physics, the Cabibbo-Kobayashi-Maskawa matrix (CKM matrix, quark mixing matrix, sometimes also called KM matrix) is an unitary matrix which contains information on the strength of flavour-changing weak decays. Technically, it specifies the mismatch of quantum states of quarks when they propagate freely and when they take part in the weak interactions. It is important in the understanding of CP-violation. A precise mathematical definition of this matrix is given in the article on the formulation of the standard model. This matrix was introduced for three generations of quarks by Makoto Kobayashi and Toshihide Maskawa, adding one generation to the matrix previously introduced by Nicola Cabibbo. This matrix is also an extension of the GIM mechanism, which only includes 2 of the 3 current families of quarks.

To proceed further, it is necessary to count the number of parameters in this matrix, V which appear in experiments, and therefore, are physically important. If there are N generations of quarks ($2N$ flavors) then

- An $N \times N$ unitary matrix (that is, a matrix V such that $VV^\dagger = I$, where $V^\dagger$ is the conjugate transpose of V and I is the identity matrix) requires N^2 real parameters to be specified.
- $2N - 1$ of these parameters are not physically significant, because one phase can be absorbed into each quark field, but an overall common phase is unobservable. Hence, the total number of free variables independent of the choice of the phases of basis vectors is $N^2 - (2N - 1) = (N - 1)^2$.
- Of these, $N(N - 1)/2$ are rotation angles called quark *mixing angles*.
- The remaining $(N - 1)(N - 2)/2$ are complex phases, which cause CP violation.

For the case $N = 2$, there is only one parameter which is a mixing angle between two generations of quarks. Historically, this was the first version of CKM matrix when only two generations were known. It is called the Cabibbo angle after its inventor Nicola Cabibbo.

For the Standard Model case ($N = 3$), there are three mixing angles and one CP-violating complex phase.

Cabibbo's idea originated from a need to explain two observed phenomena:

1. the transitions $u \leftrightarrow d$, $e \leftrightarrow v_e$, and $\mu \leftrightarrow v_\mu$ had similar amplitudes.
2. the transitions with change in strangeness $\Delta S = 1$ had amplitudes equal to 1/4 of those with $\Delta S = 0$.

Charles Kuen Kao

Charles Kuen Kao, is a pioneer in the development and use of fiber optics in telecommunications. Kao, widely regarded as the Father of Fiber Optics or Father of Fiber Optic Communications, was awarded half of the 2009 Nobel Prize in Physics for groundbreaking achievements concerning the transmission of light in fibers for optical communication.

Born	4 November 1933 (age 76), Shanghai, China
Residence	Shanghai, China (1933–1948), Hong Kong, England (1952–1970), United States
Nationality	United States of America, United Kingdom
Fields	Optics, Electrical engineering, enterprise, higher education, environmental studies, public policy, energy policy
Institutions	Chinese University of Hong Kong, ITT Corporation, Standard Telephones and Cables
Alma mater	University College London (PhD 1965, issued by University of London), Woolwich Polytechnic (BSc 1957, issued by University of London),St. Joseph's College, Hong Kong (1952)
Doctoral advisor	Harold Barlow
Known for	Fiber optics, fiber-optic communication
Notable awards	IEEE Morris N. Liebmann Memorial Award (1978), IEEE Alexander Graham Bell Medal (1985), Marconi Prize (1985), Faraday Medal (1989), James C. McGroddy Prize for New Materials (1989), Prince Philip Medal (1996) Japan Prize (1996), 3463 Kaokuen (1996), Charles Stark Draper Prize (1999), Asian of the Century (1999), Nobel Prize in Physics (2009)

FIBER-OPTIC COMMUNICATION (2009)

Fiber-optic communication is a method of transmitting information from one place to another by sending pulses of light through an optical fiber. The light forms an electromagnetic carrier wave that is modulated to carry information. First developed in the 1970s, fiber-optic communication systems have revolutionized the telecommunications industry and have played a major role in the advent of the Information Age. Because of its advantages over electrical transmission, optical fibers have largely replaced copper wire communications in core networks in the developed world.

The process of communicating using fiber-optics involves the following basic steps: creating the optical signal involving the use of a transmitter, relaying the signal along the fiber, ensuring that the signal does not become too distorted or weak, receiving the optical signal, and converting it into an electrical signal.

In 1966 Charles K. Kao and George Hockham proposed optical fibers at STC Laboratories (STL), Harlow, when they showed that the losses of 1000 db/km in existing glass (compared to 5–10 db/km in coaxial cable) was due to contaminants, which could potentially be removed.

Optical fiber was successfully developed in 1970 by Corning Glass Works, with attenuation low enough for communication purposes (about 20 dB/km), and at the same time GaAs semiconductor lasers were developed that were compact and therefore suitable for transmitting light through fiber optic cables for long distances.

After a period of research starting from 1975, the first commercial fiber-optic communications system was developed, which operated at a wavelength around 0.8 μm and used GaAs semiconductor lasers. This first-generation system operated at a bit rate of 45 Mbps with repeater spacing of up to 10 km. Soon on 22 April, 1977, General Telephone and Electronics sent the first live telephone traffic through fiber optics at a 6 Mbps throughput in Long Beach, California.

Willard Sterling Boyle

Willard Sterling Boyle, is a Canadian physicist and co-inventor of the charge-coupled device. On 6 October 2009 it was announced that he would share the 2009 Nobel Prize in Physics for the invention of an imaging semiconductor circuit—the CCD sensor.

Born	19 August 1924 (age 85), Amherst, Nova Scotia, Canada
Residence	Canada
Citizenship	Canada, United States
Fields	Applied physics
Institutions	Bell Labs
Alma mater	McGill University, Lower Canada College
Known for	Charge-coupled device
Notable awards	IEEE Morris N. Liebmann Memorial Award Draper Prize, Nobel Prize in Physics (2009)

CHARGE-COUPLED DEVICE (2009)

A charge-coupled device (CCD) is a device for the movement of electrical charge, usually from within the device to an area where the charge can be manipulated, for example conversion into a digital value. This is achieved by shifting the signals between stages within the device one at a time. Technically, CCDs are implemented as shift registers that move charge between capacitive *bins* in the device, with the shift allowing for the transfer of charge between bins.

Often the device is integrated with an image sensor, such as a photoelectric device to produce the charge that is being read, thus making the CCD a major technology for digital imaging. Although CCDs are not the only technology to allow for light detection, CCDs are widely used in professional, medical, and scientific applications where high-quality image data are required.

The charge-coupled device was invented in 1969 at AT & T Bell Labs by Willard Boyle and George E. Smith. The lab was working on semiconductor bubble memory when Boyle and Smith conceived of the design of what they termed, in their notebook, Charge Bubble Devices. A description of how the device could be used as a shift register and as a linear and area imaging devices was described in this first entry. The essence of the design was the ability to transfer charge along the surface of a semiconductor one storage capacitor to the next.

The initial paper describing the concept listed possible uses as a memory, a delay line, and an imaging device. The first experimental device demonstrating the principle was a row of closely spaced metal squares on an oxidized silicon surface electrically accessed by wire bonds.

The first working CCD made with integrated circuit technology was a simple 8-bit shift register. This device had input and output circuits and was used to demonstrate use as a shift register and as a crude eight pixel linear imaging device. Development of the device progressed at a rapid rate. By 1971, Bell researchers Michael F. Tompsett et al. were able to capture images with simple linear devices.

Several companies, including Fairchild Semiconductor, RCA and Texas Instruments, picked up on the invention and began development programs. Fairchild's effort, led by ex-Bell researcher Gil Amelio, was the first with commercial devices, and by 1974 had a linear 500-element device and a 2 D 100 × 100 pixel device. Under the leadership of Kazuo Iwama, Sony also started a big development effort on CCDs involving a significant investment. Eventually, Sony managed to mass produce CCDs for their camcorders. Before this happened, Iwama died in August 1982. Subsequently, a CCD chip was placed on his tombstone to acknowledge his contribution.

In January 2006, Boyle and Smith were awarded the National Academy of Engineering Charles Stark Draper Prize, and in 2009 they were awarded the Nobel Prize for Physics, for their work on the CCD.

George Elwood Smith

George Elwood Smith, is an American scientist, applied physicist, and co-inventor of the charge-coupled device. He was awarded a one-quarter share in the 2009 Nobel Prize in Physics for the invention of an imaging semiconductor circuit — the CCD sensor.

Born	10 May 1930 (age 80), White Plains, New York
Nationality	United States
Fields	Applied physics
Institutions	Bell Labs
Alma mater	University of Chicago, University of Pennsylvania
Known for	Charge-coupled device
Notable awards	IEEE Morris N. Liebmann Memorial Award (1974) Draper Prize (2006), Nobel Prize in Physics (2009)

CHARGE-COUPLED DEVICE (2009)

In a CCD for capturing images, there is a photoactive region, and a transmission region made out of a shift register.

An image is projected through a lens onto the capacitor array, causing each capacitor to accumulate an electric charge proportional to the light intensity at that location. A one-dimensional array, used in line-scan cameras, captures a single slice of the image, while a two-dimensional array, used in video and still cameras, captures a two-dimensional picture corresponding to the scene projected onto the focal plane of the sensor. Once the array has been exposed to the image, a control circuit causes each capacitor to transfer its contents to its neighbor. The last capacitor in the array dumps its charge into a charge amplifier, which converts the charge into a voltage. By repeating this process, the controlling circuit converts the entire contents of the array in the semiconductor to a sequence of voltages. In a digital device, these voltages are then sampled, digitized, and usually stored in memory; in an analog device, they are processed into a continuous analog signal which is then processed and fed out to other circuits for transmission, recording, or other processing.

The photoactive region of the CCD is, generally, an epitaxial layer of silicon. It has a doping of p+ (Boron) and is grown upon a substrate material, often p++. In buried channel devices, the type of design utilized in most modern CCDs, certain areas of the surface of the silicon are ion implanted with phosphorus, giving them an n-doped designation. This region defines the channel in which the photo generated charge packets will travel. The gate oxide, i.e. the capacitor dielectric, is grown on top of the epitaxial layer and substrate. Later on in the process polysilicon gates are deposited by chemical vapor deposition, patterned with photolithography, and etched in such a way that the separately phased gates lie perpendicular to the channels. The channels are further defined by utilization of the LOCOS process to produce the channel stop region. Channel stops are thermally grown oxides that serve to isolate the charge packets in one column from those in another. These channel stops are produced before the polysilicon gates are, as the LOCOS process utilizes a high temperature step that would destroy the gate material. The channels stops are parallel to, and exclusive of, the channel, or charge carrying, regions. Channel stops often have a p+ doped region underlying them, providing a further barrier to the electrons in the charge packets.

One should note that the clocking of the gates, alternately high and low, will forward and reverse bias to the diode that is provided by the buried channel (n-doped) and the epitaxial layer (p-doped). This will cause the CCD to deplete, near the p-n junction and will collect and move the charge packets beneath the gates, and within the channels of the device.

CCD manufacturing and operation can be optimized for different uses. The above process describes a frame transfer CCD. While CCDs may be manufactured on a heavily doped p++ wafer it is also possible to manufacture a device inside p-wells that have been placed on an n-wafer. This second method, reportedly, reduces smear, dark current, and infrared and red response. This method of manufacture is used in the construction of interline transfer devices.

Andre Konstant inovich Geim

Andre Konstantinovich Geim, FRS is a Dutch-Russian physicist working at the University of Manchester. Geim was awarded the 2010 Nobel Prize in Physics jointly with Konstantin Novoselov for his work on graphene. He is the Langworthy Professor and director of Manchester Centre for Mesoscience and Nanotechnology at the University of Manchester.

Born	1 October 1958 (age 52), Sochi, Russian SFSR, USSR
Residence	England
Citizenship	Netherlands
Institutions	Moscow Institute of Physics and Technology, Institute of Solid State Physics, Russian Academy of Sciences, University of Manchester, Radboud University Nijmegen
Notable students	Konstantin Novoselov
Known for	Work on graphene Levitating a frog Developing gecko tape
Notable awards	Nobel Prize (2000), Mott Prize (2007) Euro Physics Prize (2008), Körber Prize (2009) John J. Carty Award (2010), Hughes Medal (2010), Nobel Prize in Physics (2010)

GRAPHENE (2010)

Geim's achievements include the discovery of a simple method for isolating single atomic layers of graphite, known as graphene, in collaboration with researchers at the University of Manchester and IMT. The team published their findings in October 2004 in Science.

Graphene consists of one-atom-thick layers of carbon atoms arranged in two-dimensional hexagons, and is the thinnest material in the world, as well as one of the strongest and hardest. The material has many potential applications and is considered a superior alternative to silicon.

Geim said one of the first applications of graphene could be in the development of flexible touchscreens, and that he has not patented the material because he would need a specific application and an industrial partner.

In essence, graphene is an isolated atomic plane of graphite. From this perspective, graphene has been known since the invention of X-ray crystallography. Graphene planes become even better separated in intercalated graphite compounds. In 2004 physicists at the University of Manchester and the Institute for Microelectronics Technology, Chernogolovka, Russia, first isolated individual graphene planes by using Scotch tape. They also measured electronic properties of the obtained flakes and showed their unique properties. In 2005 the same Manchester Geim group together with the Philip Kim group from Columbia University demonstrated that quasiparticles in graphene were mass less Dirac fermions. These discoveries led to an explosion of interest in graphene.

Since then, hundreds of researchers have entered the area, resulting in an extensive search for relevant earlier papers. Manchester researchers themselves published the first literature review. They cite several papers in which graphene or ultra-thin graphitic layers were epitaxially grown on various substrates. Also, they note a number of reports in which intercalated graphite compounds were studied in a transmission electron microscope. In the latter case, researchers occasionally observed extremely thin graphitic flakes (few-layer graphene and possibly even individual layers). An early detailed study on few-layer graphene dates back to 1962. The earliest TEM images of few-layer graphene were published by G. Ruess and F. Vogt in 1948. However, already D.C. Brodie was aware of the highly lamellar structure of thermally reduced graphite oxide in 1859. It was studied in detail by V. Kohlschütter and P. Haenni in 1918, who also described the properties of graphite oxide paper.

Konstantin Sergeevich Novoselov

Konstantin Sergeevich Novoselov is a Russo-British physicist, most notably known for his works on graphene together with Andre Geim, which earned them the Nobel Prize in Physics in 2010. Novoselov is currently a member of the mesoscopic research group at the University of Manchester as a Royal Society University Research Fellow. Novoselov is also recipient of an ERC Starting Grant from the European Research Council.

Born	23 August 1974 (age 36), Nizhny Tagil, Russian SFSR, USSR
Residence	England
Citizenship	Russia and United Kingdom
Nationality	Russian
Institutions	University of Manchester
Alma mater	Moscow Institute of Physics and Technology, University of Nijmegen
Doctoral advisor	Jan Kees Maan, Andre Geim
Known for	Study of graphene
Notable awards	Nobel Prize in Physics (2010)

GRAPHENE (2010)

Konstantin Novoselov was born in Nizhny Tagil, Soviet Union, in 1974 in a Russian family. He received a Diploma from the Moscow Institute of Physics and Technology, and undertook his PhD studies at the University of Nijmegen in the Netherlands before moving to the University of Manchester in the United Kingdom with his doctoral advisor Andre Geim in 2001. He now holds both Russian and British citizenship.

Novoselov has published more than 60 peer-reviewed research papers, on topics like mesoscopic superconductivity (Hall magnetometry), sub-atomic movements of magnetic domain walls, the invention of gecko tape, and graphene.

Graphene is an allotrope of carbon, whose structure is one-atom-thick planar sheets of sp2-bonded carbon atoms that are densely packed in a honeycomb crystal lattice. The term graphene was coined as a combination of graphite and the suffix -ene by Hanns-Peter Boehm, who described single-layer carbon foils in 1962. Graphene is most easily visualized as an atomic-scale chicken wire made of carbon atoms and their bonds. The crystalline or flake form of graphite consists of many graphene sheets stacked together.

The carbon-carbon bond length in graphene is about 0.142 nanometers. Graphene sheets stack to form graphite with an inter planar spacing of 0.335 nm, which means that a stack of 3 million sheets would be only one millimeter thick. Graphene is the basic structural element of some carbon allotropes including graphite, charcoal, carbon nano tubes and fullerenes. It can also be considered as an indefinitely large aromatic molecule, the limiting case of the family of flat polycyclic aromatic hydrocarbons.

The Nobel Prize in Physics for 2010 was awarded to Andre Geim and Konstantin Novoselov for groundbreaking experiments regarding the two-dimensional material graphene.

Saul Perlmutter

Saul Perlmutter (born 1959) is an American astrophysicist at the Lawrence Berkeley National Laboratory and a professor of physics at the University of California, Berkeley. He is a member of the American Academy of Arts and Sciences, and was elected a Fellow of the American Association for the Advancement of Science in 2003. He is also a member of the National Academy of Sciences. Perlmutter shared both the 2006 Shaw Prize in Astronomy and the 2011 Nobel Prize in Physics with Adam Riess and Brian P. Schmidt for providing evidence that the expansion of the universe is accelerating.

Born	1959 (age 51–52) Champaign-Urbana, Illinois, U.S.
Residence	United States
Nationality	American
Fields	Physics
Institutions	UC Berkeley/LBNL
Alma mater	Harvard (AB) / UC Berkeley (PhD)
Doctoral advisor	Richard A. Muller
Known for	Accelerating universe/Dark energy
Notable awards	Ernest Orlando Lawrence Award (2002), Shaw Prize in Astronomy (2006), Gruber Prize in Cosmology (2007), Nobel Prize in Physics (2011)

ACCELERATING UNIVERSE

The accelerating universe is the observation that the universe appears to be expanding at an increasing rate, which in formal terms means that the cosmic scale factor a (t) has a positive second derivative, implying that the velocity at which a given galaxy is receding from us should be continually increasing over time (here the recession velocity is the same one that appears in Hubble's law; defining 'velocity' in cosmology is somewhat subtle. In 1998, observations of type Ia supernovae suggested that the expansion of the universe has been accelerating since around redshift of z~0.5. The 2006 Shaw Prize in Astronomy and the 2011 Nobel Prize in Physics were both awarded to Saul Perlmutter, Brian P. Schmidt, and Adam G. Riess for the 1998 discovery of the accelerating expansion of the Universe through observations of distant supernova.

After the initial discovery in 1998, these observations were corroborated by several independent sources: the cosmic microwave background radiation and large scale structure, apparent size of baryon acoustic oscillations, age of the universe, as well as improved measurements of the supernova, and X-ray properties of galaxy clusters.

An expanding universe means that density drops due to continual space being added between all matter. If acceleration continues, eventually all galaxies beyond our own local super cluster will redshift so far that it will become hard to detect them, and the distant universe will turn dark.

Models attempting to explain accelerating expansion include some form of dark energy, cosmological constant, quintessence, dark fluid or phantom energy. The most important property of dark energy is that it has negative pressure which is distributed relatively homogeneously in space.

Phantom energy in a scenario known as the Big Rip causes an exponentially increasing divergent expansion, which overcomes the gravitation of the local group and tears apart our Virgo super cluster; it then tears apart the Milky Way Galaxy, our solar system, and finally even atoms. Measurements of acceleration are thought crucial to determining the ultimate fate of the universe, however we should expect the implications of such a major discovery to develop slowly over many years in the same way the Big Bang model has continued to develop.

Brian Schmidt

Brian P. Schmidt (born February 24, 1967) is a Distinguished Professor, Australian Research Council Laureate Fellow and astrophysicist at the Australian National University Mount Stromlo Observatory and Research School of Astronomy and Astrophysics and is widely known for his research in using supernovae as Cosmological Probes. He currently holds an Australia Research Council Federation Fellowship. Schmidt shared both the 2006 Shaw Prize in Astronomy and the 2011 Nobel Prize in Physics with Saul Perlmutter and Adam Riess for providing evidence that the expansion of the universe is accelerating.

Born	February 24, 1967 (age 44), Missoula, Montana, United States
Alma mater	University of Arizona (1989), Harvard University (1993)
Doctoral advisor	Robert Kirshner
Notable awards	Shaw Prize in Astronomy (2006) Nobel Prize in Physics (2011)

DARK ENERGY

In physical cosmology, astronomy and celestial mechanics, dark energy is a hypothetical form of energy that permeates all of space and tends to increase the rate of expansion of the universe. Dark energy is the most accepted theory to explain recent observations that the universe appears to be expanding at an accelerating rate. In the standard model of cosmology, dark energy currently accounts for 73% of the total mass-energy of the universe.

Two proposed forms for dark energy are the cosmological constant, a constant energy density filling space homogeneously, and scalar fields such as quintessence or moduli, dynamic quantities whose energy density can vary in time and space. Contributions from scalar fields that are constant in space are usually also included in the cosmological constant. The cosmological constant is physically equivalent to vacuum energy. Scalar fields which do change in space can be difficult to distinguish from a cosmological constant because the change may be extremely slow.

High-precision measurements of the expansion of the universe are required to understand how the expansion rate changes over time. In general relativity, the evolution of the expansion rate is parameterized by the cosmological equation of state (the relationship between temperature, pressure, and combined matter, energy, and vacuum energy density for any region of space). Measuring the equation of state for dark energy is one of the biggest efforts in observational cosmology today.

Adding the cosmological constant to cosmology's standard FLRW metric leads to the Lambda-CDM model, which has been referred to as the standard model of cosmology because of its precise agreement with observations. Dark energy has been used as a crucial ingredient in a recent attempt to formulate a cyclic model for the universe.

A 2011 survey of more than 200,000 galaxies appears to confirm the existence of dark energy, although the exact physics behind it remains unknown.

Adam Riess

Adam Guy Riess (born December 1969, Washington, D.C.) is an American astrophysicist at Johns Hopkins University and the Space Telescope Science Institute and is widely known for his research in using supernovae as Cosmological Probes. Riess shared both the 2006 Shaw Prize in Astronomy and the 2011 Nobel Prize in Physics with Saul Perlmutter and Brian P. Schmidt for providing evidence that the expansion of the universe is accelerating.

Born	December 1969 (age 41), Washington, D.C., U.S.
Residence	United States
Nationality	American
Fields	Physics
Institutions	Johns Hopkins University/Space Telescope Science Institute
Alma mater	Massachusetts Institute of Technology, Harvard University, University of California, Berkeley
Known for	Accelerating universe/Dark energy
Notable awards	Shaw Prize in Astronomy (2006) Nobel Prize in Physics (2011)

NATURE OF DARK ENERGY

The nature of this dark energy is a matter of speculation. The evidence for dark energy is only indirect coming from distance measurements and their relation to redshift. It is thought to be very homogeneous, not very dense and is not known to interact through any of the fundamental forces other than gravity. Since it is not very dense—roughly 10^{-29} grams per cubic centimeter—it is hard to imagine experiments to detect it in the laboratory. Dark energy can only have such a profound impact on the universe, making up 74% of universal density, because it uniformly fills otherwise empty space. The two leading models are a cosmological constant and quintessence. Both models include the common characteristic that dark energy must have negative pressure.

NEGATIVE PRESSURE

Independently from its actual nature, dark energy would need to have a strong negative pressure (acting repulsively) in order to explain the observed acceleration in the expansion rate of the universe.

According to general relativity, the pressure within a substance contributes to its gravitational attraction for other things just as its mass density does. This happens because the physical quantity that causes matter to generate gravitational effects is the Stress-energy tensor, which contains both the energy (or matter) density of a substance and its pressure and viscosity.

In the Friedmann-Lemaître-Robertson-Walker metric, it can be shown that a strong constant negative pressure in the entire universe causes acceleration in universe expansion, if the universe is already expanding or a deceleration in universe contraction, if the universe is already contracting. More exactly, the second derivative of the universe scale factor, $\ddot{a}$, is positive if the equation of state of the universe is such that $w < -1/3$

This accelerating expansion effect is sometimes labeled gravitational repulsion, which is a colorful but possibly confusing expression. In fact a negative pressure does not influence the gravitational interaction between masses, which remains attractive—but rather alters the overall evolution of the universe at the cosmological scale, typically resulting in the accelerating expansion of the universe despite the attraction among the masses present in the universe.

The first suggestion for dark energy from observed data happened in 1992, when György Paál and his collaborators published a paper. Previously, in 1990 Broadhurst et al. had published the so called "pencil beam survey" about the irregularities in the galaxy distribution. Using this data Paál et al. found in some cosmological model the irregularities became more regular. In these models the cosmological constant (dark energy) was needed. Two years later in another paper they suggested that $\Omega_A = 2/3$ later observations confirmed 'this value.

IMPORTANT FACTS ABOUT NOBEL PRIZE

The word "Laureate" refers to being signified by the laurel wreath. In Greek mythology, the god Apollo is represented wearing a laurel wreath on his head. A laurel wreath is a circular crown made of branches and leaves of the bay laurel (In latin: Laurus nobilis). In ancient Greek laurel wreaths were awarded to victors as a sign of honour—both in athletic competitions and in poetic meets.

Number of Nobel Prizes in Physics

104 Nobel Prizes in Physics have been awarded since 1901. It was not awarded on six occasions: in 1916, 1931, 1934, 1940, 1941, and 1942.

In the statutes of the Nobel Foundation it says: "If none of the works under consideration is found to be of the importance indicated in the first paragraph, the prize money shall be reserved until the following year. If, even then, the prize cannot be awarded, the amount shall be added to the Foundation's restricted funds." During World War I and II, fewer Nobel Prizes were awarded.

Number of Shared and Unshared Nobel Prizes in Physics

47 Physics Prizes have been given to one Laureate only. 29 Physics Prizes have been shared by two Laureates. 28 Physics Prizes have been shared between three Laureates.

In the statutes of the Nobel Foundation it says: A prize amount may be equally divided between two works, each of which is considered to merit a prize. If a work that is being rewarded has been produced by two or three persons, the prize shall be awarded to them jointly. In no case may a prize amount be divided between more than three persons.

Number of Nobel Laureates in Physics

The Nobel Prize in Physics has been awarded to 189 Laureates. As John Bardeen has been awarded twice there are 188 individuals who have been awarded the Nobel Prize in Physics since 1901.

Youngest Physics Laureate

To date, the youngest Nobel Laureate in Physics is Lawrence Bragg, who was 25 years old when he was awarded the Nobel

Prize with his father in 1915. Bragg is not only the youngest Physics Laureate, he is also the youngest Nobel Laureate in any Nobel Prize area.

Oldest Physics Laureate

The oldest Nobel Laureate in Physics till date is Raymond Davis Jr., who was 88 years old when he was awarded the Nobel Prize in 2002.

Female Nobel Laureates in Physics

Of the 188 individuals awarded the Nobel Prize in Physics, only two are women.

1903–Marie Curie (also awarded the 1911 Nobel Prize in Chemistry).

1963–Maria Goeppert-Mayer.

Multiple Nobel Laureates in Physics.

John Bardeen is the only person who has received the Nobel Prize in Physics twice. Marie Curie was awarded the Nobel Prize twice, once in Physics and once in Chemistry.

➤ John Bardeen: 1956 and 1972.

➤ Marie Curie: 1903 (Physics) and 1911 (Chemistry).

Posthumous Nobel Prizes in Physics

There have been no posthumous Nobel Prizes in Physics. From 1974, the Statutes of the Nobel Foundation stipulate that a Prize cannot be awarded posthumously, unless death has occurred after the announcement of the Nobel Prize. Before 1974, the Nobel Prize has only been awarded posthumously twice: to Dag Hammarskjöld (Nobel Peace Prize 1961) and Erik Axel Karlfeldt (Nobel Prize in Literature 1931).

Family Nobel Laureates in Physics
Married couples

Marie Curie and Pierre Curie were awarded the Nobel Prize in Physics in 1903. Marie Curie was awarded the Nobel Prize second time in 1911, this time receiving the Nobel Prize in Chemistry.

(One of Marie and Pierre Curie's daughters, Irène Joliot-Curie, was awarded the Nobel Prize in Chemistry in 1935 together with her husband Frédéric Joliot.)

Father and son

1. William Bragg and Lawrence Bragg, 1915
2. Niels Bohr, 1922 and Aage N. Bohr, 1975
3. Manne Siegbahn, 1924 and Kai M. Siegbahn, 1981
4. J. J. Thomson, 1906 and George Paget Thomson, 1937

The Nobel Prize amount for 2009 is set at Swedish kronor (SEK) 10 million per full Nobel Prize.

The Nobel Medal for Physics and Chemistry

Registered trademark of the Nobel Foundation

The medal of The Royal Swedish Academy of Sciences represents Nature in the form of a goddess resembling Isis, emerging from the clouds and holding in her arms a cornucopia. The veil which covers her cold and austere face is held up by the Genius of Science.

The inscription reads

Inventas vitam juvat excoluisse per artes:

loosely translated "And they who bettered life on earth by their newly found mastery." (Word for word: inventions enhance life which is beautified through art).

The words are taken from Vergilius Aeneid, the 6th song, verse 663:

Lo, God-loved poets, men who spake things worthy Phoebus' heart; and they who bettered life on earth by new-found mastery.

The name of the Nobel Laureate is engraved on the plate below the figures, and the text "REG. ACAD. SCIENT. SUEC." stands for The Royal Swedish Academy of Sciences.

The Nobel Medal for Physics and Chemistry was designed by Erik Lindberg.

ALL NOBEL PRIZES IN PHYSICS

The Nobel Prize in Physics has been awarded 104 times to 188 Nobel Laureates between 1901 and 2010. John Bardeen is the only Nobel Laureate who has been awarded the Nobel Prize in Physics twice, in 1956 and 1972. This means that a total of 188 individuals have received the Nobel Prize in Physics.

2010:　Andre Geim, konstantin Novoselov

2009:　Charles K. Kao, Willard S. Boyle, George E. Smith

2008:　Yoichiro Nambu, Makoto Kobayashi, Toshihide Maskawa

2007:　Albert Fert, Peter Grünberg

2006:　John C. Mather, George F. Smoot

2005:　Roy J. Glauber, John L. Hall, Theodor W. Hänsch

2004:　David J. Gross, H. David Politzer, Frank Wilczek

2003:　Alexei A. Abrikosov, Vitaly L. Ginzburg, Anthony J. Leggett

2002:　Raymond Davis Jr., Masatoshi Koshiba, Riccardo Giacconi

2001:　Eric A. Cornell, Wolfgang Ketterle, Carl E. Wieman

2000:　Zhores I. Alferov, Herbert Kroemer, Jack S. Kilby

1999:　Gerardus 't Hooft, Martinus J.G. Veltman

1998:　Robert B. Laughlin, Horst L. Störmer, Daniel C. Tsui

1997:　Steven Chu, Claude Cohen-Tannoudji, William D. Phillips

1996:　David M. Lee, Douglas D. Osheroff, Robert C. Richardson

1995:　Martin L. Perl, Frederick Reines

1994:　Bertram N. Brockhouse, Clifford G. Shull

1993:　Russell A. Hulse, Joseph H. Taylor Jr.

1992:　Georges Charpak

1991: Pierre-Gilles de Gennes

1990: Jerome I. Friedman, Henry W. Kendall, Richard E. Taylor

1989: Norman F. Ramsey, Hans G. Dehmelt, Wolfgang Paul

1988: Leon M. Lederman, Melvin Schwartz, Jack Steinberger

1987: J. Georg Bednorz, K. Alexander Müller

1986: Ernst Ruska, Gerd Binnig, Heinrich Rohrer

1985: Klaus von Klitzing

1984: Carlo Rubbia, Simon van der Meer

1983: Subramanyan Chandrasekhar, William Alfred Fowler

1982: Kenneth G. Wilson

1981: Nicolaas Bloembergen, Arthur Leonard Schawlow, Kai M. Siegbahn

1980: James Watson Cronin, Val Logsdon Fitch

1979: Sheldon Lee Glashow, Abdus Salam, Steven Weinberg

1978: Pyotr Leonidovich Kapitsa, Arno Allan Penzias, Robert Woodrow Wilson

1977: Philip Warren Anderson, Sir Nevill Francis Mott, John Hasbrouck van Vleck

1976: Burton Richter, Samuel Chao Chung Ting

1975: Aage Niels Bohr, Ben Roy Mottelson, Leo James Rainwater

1974: Sir Martin Ryle, Antony Hewish

1973: Leo Esaki, Ivar Giaever, Brian David Josephson

1972: John Bardeen, Leon Neil Cooper, John Robert Schrieffer

1971: Dennis Gabor

1970: Hannes Olof Gösta Alfvén, Louis Eugène Félix Néel

1969: Murray Gell-Mann

1968: Luis Walter Alvarez

1967: Hans Albrecht Bethe

1966: Alfred Kastler

1965: Sin-Itiro Tomonaga, Julian Schwinger, Richard P. Feynman

1964: Charles Hard Townes, Nicolay Gennadiyevich Basov, Alexander Mikhaylovich Prokhorov

1963: Eugene Paul Wigner, Maria Goeppert-Mayer, J. Hans D. Jensen

1962: Lev Davidovich Landau

1961: Robert Hofstadter, Rudolf Ludwig Mössbauer

1960: Donald Arthur Glaser

1959: Emilio Gino Segrè, Owen Chamberlain

1958: Pavel Alekseyevich Cherenkov, Il'ja Mikhailovich Frank, Igor Yevgenyevich Tamm

1957: Chen Ning Yang, Tsung-Dao Lee

1956: William Bradford Shockley, John Bardeen, Walter Houser Brattain

1955: Willis Eugene Lamb, Polykarp Kusch

1954: Max Born, Walther Bothe

1953: Frits (Frederik) Zernike

1952: Felix Bloch, Edward Mills Purcell

1951: Sir John Douglas Cockcroft, Ernest Thomas Sinton Walton

1950: Cecil Frank Powell

1949: Hideki Yukawa

1948: Patrick Maynard Stuart Blackett

1947: Sir Edward Victor Appleton

1946: Percy Williams Bridgman

1945: Wolfgang Pauli

1944: Isidor Isaac Rabi

1943: Otto Stern

1939: Ernest Orlando Lawrence

1938: Enrico Fermi

1937: Clinton Joseph Davisson, George Paget Thomson

1936: Victor Franz Hess, Carl David Anderson

1935: James Chadwick

1933: Erwin Schrödinger, Paul Adrien Maurice Dirac

1932: Werner Karl Heisenberg

1930: Sir Chandrasekhara Venkata Raman

1929: Prince Louis-Victor Pierre Raymond de Broglie

1928: Owen Willans Richardson

1927: Arthur Holly Compton, Charles Thomson Rees Wilson

1926: Jean Baptiste Perrin

1925: James Franck, Gustav Ludwig Hertz

1924: Karl Manne Georg Siegbahn

1923: Robert Andrews Millikan

1922: Niels Henrik David Bohr

1921: Albert Einstein

1920: Charles Edouard Guillaume

1919: Johannes Stark

1918: Max Karl Ernst Ludwig Planck

1917: Charles Glover Barkla

1915: Sir William Henry Bragg, William Lawrence Bragg

1914: Max von Laue

1913: Heike Kamerlingh Onnes

1912: Nils Gustaf Dalén

1911: Wilhelm Wien

1910: Johannes Diderik van der Waals

1909: Guglielmo Marconi, Karl Ferdinand Braun
1908: Gabriel Lippmann
1907: Albert Abraham Michelson
1906: Joseph John Thomson
1905: Philipp Eduard Anton von Lenard
1904: Lord Rayleigh (John William Strutt)
1903: Antoine Henri Becquerel, Pierre Curie, Marie Curie, née Sklodowska
1902: Hendrik Antoon Lorentz, Pieter Zeeman
1901: Wilhelm Conrad Röntgen

Index

Reader's Note

Reader's Note